BEYOND EXTINCTION

For Terri

BEYOND EXTINCTION: THE ETERNAL OCEAN

First Edition June 2019.

Designed by et al design consultants, etal-design.com

Printed in China by 1010 Printing (UK) Ltd, 1010 Printing.com

Published in the United Kingdom by At One Communications, AtOne.org

ISBN 978-1-9160394-0-7 (Classic Edition)
ISBN 978-1-9160394-1-4 (Boxed and signed Collectors' Edition, limited to 100 copies)

Beyond Extinction

THE ETERNAL OCEAN

Climate change & the continuity of life

WOLFGANG GRULKE

@one

The ocean is the womb of all life on earth. It is beautiful and bizarre, violent and mysterious. Inhabited by a cast of characters stolen from fantasy, it's a dystopian world where dragons are real and monsters are commonplace.

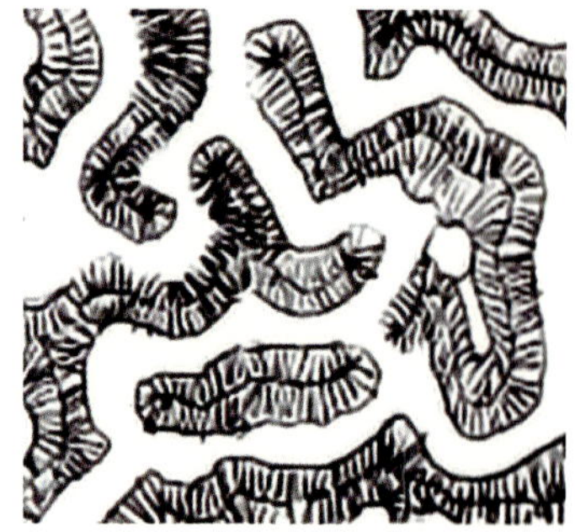

Today's spectacular marine life has an ancient history preserved in stone – fossil strata that read like dramatic pages from the longest story ever told: tales of evolution, extinction, and surprising continuity.

Having thrived a tumultuous 500 million years, this marine kingdom is now challenged by a new, arrogant and domineering life form.

This book looks beyond the media focus on climate change and extinction to celebrate the continuity of ocean life. I'll take you on a personal journey to explore origins and destinies, from primordial soup to today's threatened oceans – towards a future we can influence. We always have a choice.

OTHER BOOKS BY WOLFGANG GRULKE

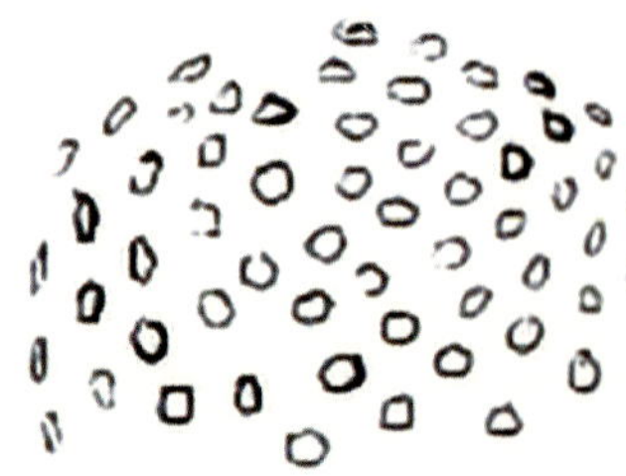

Published by At One Communications:
HETEROMORPH *The rarest fossil ammonites. Nature at its most bizarre*
NAUTILUS *Beautiful survivor. 500 million years of evolutionary history*

Published by The Financial Times/Pearson:
LESSONS IN RADICAL INNOVATION *Out of the box, straight to the bottom line*
10 LESSONS FROM THE FUTURE *Tomorrow is a matter of choice*
Spanish, Chinese and Arabic editions also available

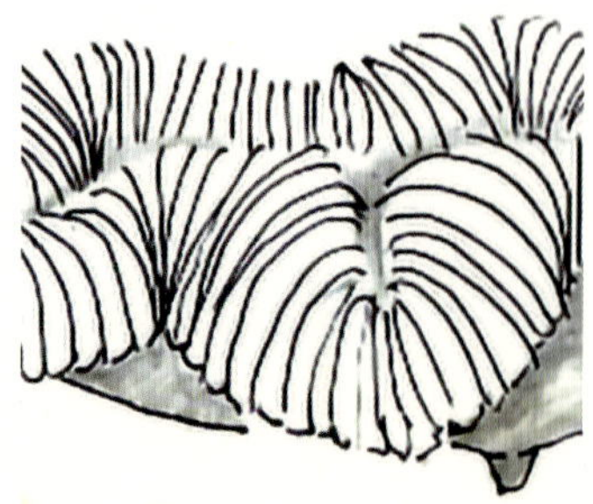

Published by Camerapix:
ALDABRA *World Heritage Site (Section on Marine Biology by Wolfgang Grulke)*

CONTENTS

As earth cooled, 4.4 billion years ago, magma forced its way from the core to the surface and an explosion of steam formed the first oceans. It rained incessantly for about a million years.

The earth's mantle still contains 80 times the amount of water in the ocean.

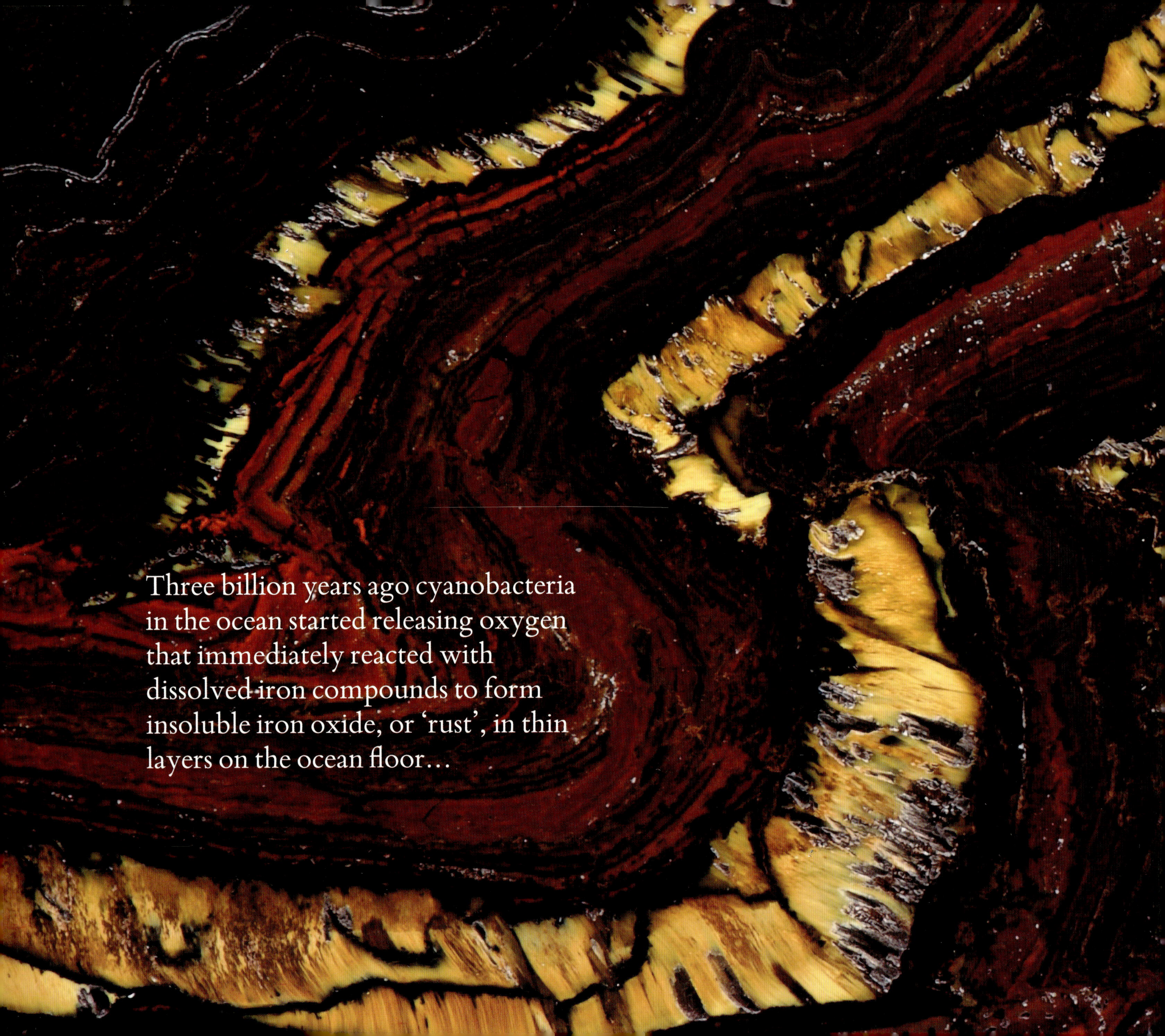

Three billion years ago cyanobacteria in the ocean started releasing oxygen that immediately reacted with dissolved iron compounds to form insoluble iron oxide, or 'rust', in thin layers on the ocean floor…

…creating these Banded Ironstone Formations.

Under a pink sky and in green seas cyanobacteria invented photosynthesis.

Over the next two billion years stromatolites began to fill the atmosphere with oxygen, prompting a massive environmental disaster for existing life forms, all of which were anaerobic…

… but making the planet habitable for aerobic life, and eventually us.

The coelacanth evolved around 400 million years ago…

…and still lives in several deep ocean habitats today.

Today, selective breeding is creating spectacular new fish forms that rely entirely on us for their survival.

Brittle stars were prolific 185 million years ago…

…and still thrive in all oceans today.

When you dive on a coral reef today…

First modern corals 250 million years ago.

First sponges 750 million years ago.

First feather stars more than 200 million years ago.

…you may as well be diving in a Jurassic ocean.

First gorgonians 200 million years ago.

PREFACE

By Paul Callomon

We have no way of knowing when humans first spotted fossils, but it must have been a very long time ago. In many places their regular shapes stand out so clearly in the surrounding rock as to catch even the uncurious eye. Recognition is easy as many fossils closely resemble things already around us, like shells, ferns – and bones.

In the great human exercise of naming things, bones have a particular significance. Animals that have them are more important than those that do not, because guessing wrong about them can have serious consequences. Correctly identified, their owners either may be avoided (tigers, lions and crocodiles) or ridden, harnessed or eaten (more or less anything else). The discovery of a bone, therefore, triggers vital questions and enhances the social authority of anyone who can reliably answer them.

Grey areas remain, however. The lore of many cultures invokes mythical beasts that were long presumed real and even a few for whom that remains the case. To dismiss as non-existent the Himalayan Yeti, the Kappa of Japan and the dragons of many lands would mean contradicting the vivid accounts of eyewitnesses. Sailors' wild-eyed tales of the Kraken persisted into fairly recent times, and when, in 1803 President Thomas Jefferson showed outbound explorer Meriwether Lewis the fresh fossil bones of giant sloths and mammoths from Kentucky, both men thought there was a good chance that the animals themselves still roamed somewhere in the vast American West.

During the great blooming of Palaeontology in the 19th century, however, the distinction between extant and extinct became steadily clearer. Though hailing from different eras, dinosaurs, ichthyosaurs and trilobites could all be firmly consigned to "prehistory". As expeditions reached every corner of the Earth it became evident that even mammoths and sabre-tooth tigers, whose bodies were barely cold in geological terms, were no longer to be found alive. Clearly, they had once flourished; equally clearly, they were now gone.

Sea shells have been treasured since early humans first decided that the shell was more than just detritus after a meal and wondered at the spiral shape of its construction. The mollusk collection at the Academy of Natural Sciences in Philadelphia is curated by a team including Paul Callomon and contains more than ten million specimens.

In building museums and writing about fossils in the Victorian era it was scientifically expedient and politically prudent to maintain a separation between the "prehistoric world" and that of today. That way, landmarks of world history such as the Creation and the Great Flood could be diplomatically skirted. Pioneering evolutionists deftly juggled established "truth" and new "knowledge" by allowing both to be incomplete and by avoiding situations where one would have to be chosen over the other.

A "fact" is what remains when controversy ceases.

Bruno Latour (paraphrased)

While it was easy to distinguish fossil vertebrates such as dinosaurs and ichthyosaurs from their modern counterparts, however, the same was not always the case with invertebrates. The petrified starfish in the Jurassic rocks of England seemed little different from their creeping cousins on the seabed just a few miles away, while the fossil Heart Urchins in the chalk cliffs of England's south coast were virtually indistinguishable from living ones that washed up on the beaches below.

Growing exploration of the deep ocean in the 19th century then revealed that groups that had up to then been thought extinct, such as the nautiluses, crinoids, slit shells and brachiopods, were still very much around. The apparently inconsequential nature of these marine animals allowed scholars to avoid mentioning the obvious truth of their continuity even as they were writing ever-more sophisticated accounts of the natural world.

The truth will out, however, and by the early 20th century scientists had demonstrated that although each species eventually becomes extinct, the higher groups of marine invertebrates almost never have. Two well-known exceptions are the ammonites and the trilobites, whose abundance in many rock layers worldwide reflects both their longevity and the major roles they played in the marine ecosystem of their times. Despite the persistent hopes of some (including me), however, both are likely gone for good.

There is no term that means "the study of all life over time – past, present and future".

In setting out the themes of this book, Wolfgang and I canvassed our academic and social networks for a term that means "the study of all life over time – past, present and future". As he explains, there seems not to be one. The fields 'biology' or 'palaeontology' cover only parts of the spectrum. I rather liked "biosophy" – the enactment of a love of living things – but that still didn't seem all-encompassing. Perhaps the condition of being so fully and timelessly attuned to nature that there is no separation between one's perception and the truth itself must by definition be un-nameable, as the Taoists first proposed many centuries ago.

In the pages that follow you will encounter lineages of animals that cruised through the traditional barriers separating named ages, seamlessly linking the distant past to the vibrant present. Interpreting natural history through blood lines over time, rather than segmented into epochs, best enhances our understanding of life on earth. It creates a more sophisticated portrait of life over time and serves to put the human race – a bunch of rude and disruptive newcomers, to put it charitably – in correct perspective.

Interpreting natural history through blood lines over time, rather than segmented into epochs, creates a more sophisticated portrait of life on earth.

That is exactly what this book attempts to do, and it ends with a message that is neither comforting nor uncontroversial. Like the Sorcerer's Apprentice, humankind has in the last few centuries carelessly set in motion processes whose immediate effects are invisible but whose collective force is likely to prove irresistible. The ingenuity that replaced buggies with automobiles and steamships with jet airliners in a single generation ran some way ahead of a deeper wisdom that is only now sounding alarms when it already might be too late.

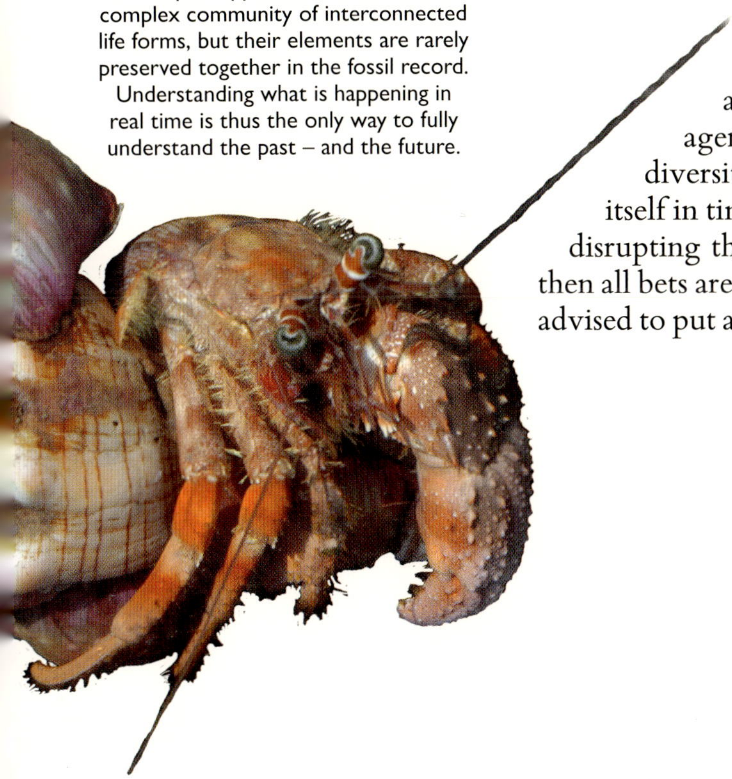

This hermit crab *Dardanus pedunculatus* and its companion sea anemones *Calliactys polypus* use a discarded sea snail shell *Pleuroploca trapezium* as a shared home. The crab's messy eating habits feed the anemones with detritus and they in turn protect it with their thread-like stinging cells. Symbiotic relationships such as this help us appreciate the ocean as a complex community of interconnected life forms, but their elements are rarely preserved together in the fossil record. Understanding what is happening in real time is thus the only way to fully understand the past – and the future.

Nevertheless, even if the charismatic megafauna – the tigers, elephants and, yes, humans – are wiped out by our agency, then the fossil record tells us that the diversity of life everywhere will happily restore itself in time. If, on the other hand, we succeed in disrupting the basic metabolism of the ocean itself, then all bets are off and shrewd investors would be best advised to put all their money into bacteria.

The myriad crustaceans, corals, echinoderms, worms and molluscs that shared the ancient seabed, remain today, their vigour and diversity utterly undiminished.

Life within the ocean evolved
three billion years before life on land.

The ocean represents 90% of earth's biosphere – it is earth's largest habitat.

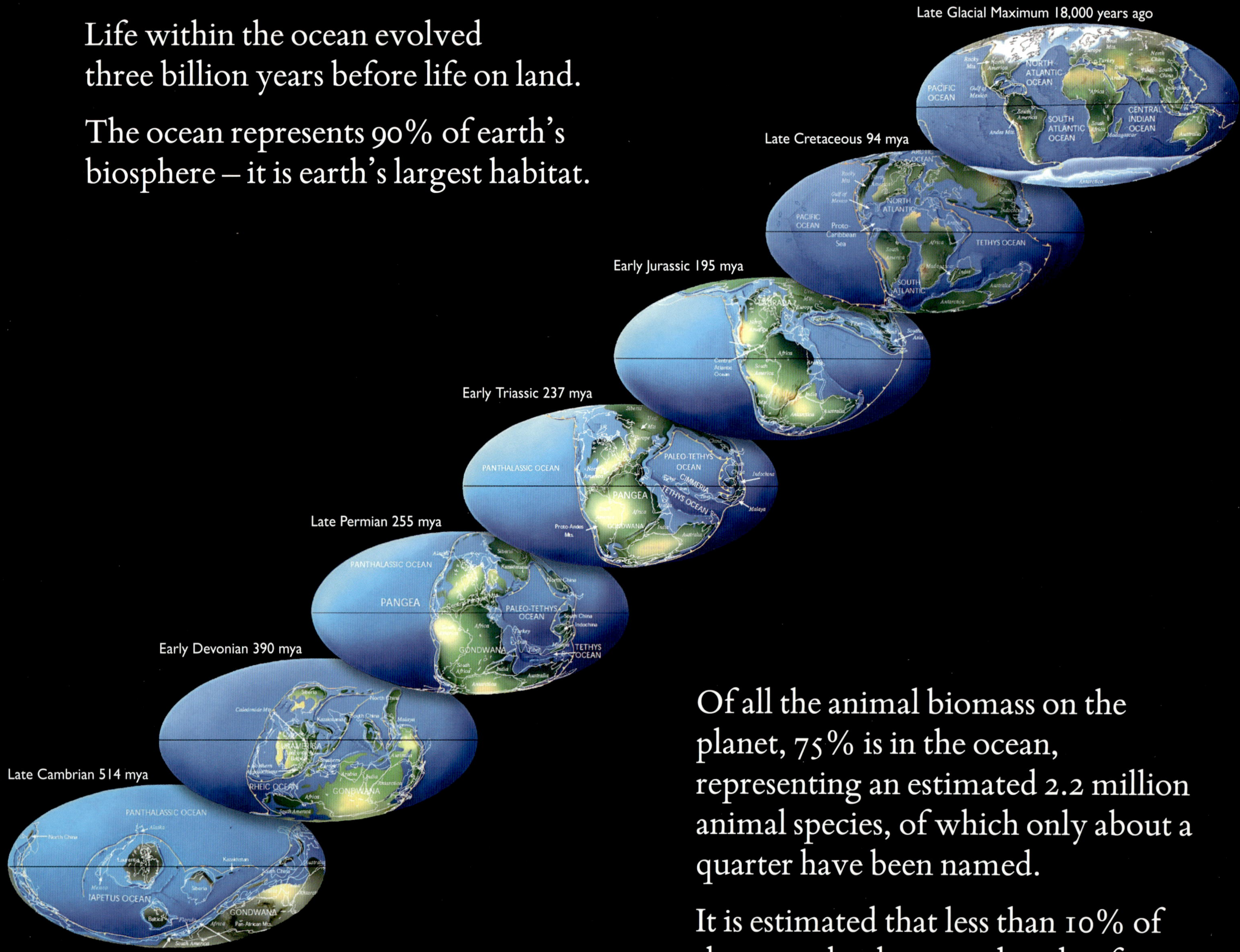

Of all the animal biomass on the planet, 75% is in the ocean, representing an estimated 2.2 million animal species, of which only about a quarter have been named.

It is estimated that less than 10% of the ocean has been explored so far.

INTRODUCTION

The ocean has existed on earth in one form or another for more than four billion years. The human concept of 'a global ocean' dates back to long before we even knew the world was round. In classical antiquity the Greeks and Romans believed in *Oceanus*, the divine personification of the sea, an enormous river encircling the world. This was a time when only the Mediterranean and the Atlantic were known to them. It was believed that gods and spirits ruled the seas and provided protection from a mysterious and dystopian underwater world.

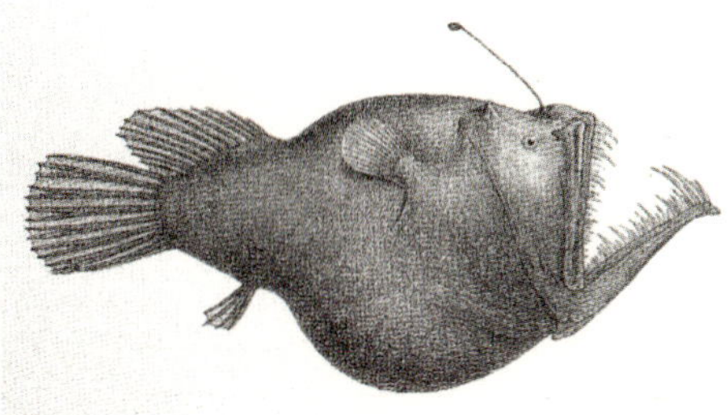

An enigmatic deep-sea fish *Melanocetus murrayi* illustrated in the Challenger Expedition Reports 1872–1876.

We are now familiar with photographs of the earth from space and know for a fact that we live on a 'blue planet'. It is self-evident that the ocean covers most of the planet, 71% to be exact. It is deeper in places than the highest mountains on earth – the Mariana Trench drops down to about 11,000 meters while Everest is merely 8,848 meters high. The mid-ocean ridge in the north Atlantic forms the longest mountain chain on earth.

Despite five mass extinctions, and many minor ones, ocean life thrives in surprising continuity.

In terms of life on earth, the ocean represents 90% of the earth's biosphere – it is earth's largest habitat by far. More than half a million animal species living in it have been named but the actual number that exist there is obviously much larger, possibly quite innumerable.

The ocean has always had a dominant effect on global climate: its ocean currents govern the world's weather, oceanic evaporation, as part of the water cycle, is the source of most rainfall, and ocean temperatures determine climate and wind patterns that affect life on land.

Steam in the atmosphere cooled enough for it to rain incessantly for about a million years.

Andrew Revkin, 2018,
Weather: An Illustrated History

As the earth cooled down from its molten origin 4.5 billion years ago, water vapour condensed and formed the oceans within half a billion years. The first simple unicellular life forms evolved there 3.5 billion years ago, fully three billion years before life advanced from the ocean onto land. The ocean literally is the womb of all life on earth.

Phytoplankton in the ocean is responsible for the creation of more than half of atmospheric oxygen.

Many land animals, all of which had their early origins in the ocean, have adapted again to living a major part of their life in it. The giant terrors of Jurassic and Cretaceous oceans, the ichthyosaurs and mosasaurs, and today's whales and dolphins all evolved from land-based reptiles, lizards and mammals. It is quite ironic to think that most of the largest animals and top predators in the ocean have evolved, quite independently, from forms that previously lived on land. More about that later.

There have indeed been several relatively sudden, widespread and truly dramatic extinction events that can be used as convenient boundaries in time, but it is important to realise that they affected the various forms of life on earth to widely differing degrees and some effectively not at all. The end of the dinosaurs was not the end of the crinoids, for example, and the still mystifying demise of the trilobites after more than 200 million years of success did not apparently bother the nautiluses one bit.

A personal journey – from extinction to continuities

Based on your feedback from the first two books in this trilogy, *Heteromorph* and *Nautilus*, you convinced me to make this final instalment a deeply personal journey.

The ocean has kept me perpetually entranced. For as long as I can remember I have been passionately curious, an avid traveller and a collector of beautiful things, images and ideas. I love to connect and share as I learn. These traits have shaped me and my relationships with nature.

This is not a comprehensive text book on the life forms that inhabit the ocean: think of it rather as their story told from a selection of my own experiences. I'll share photos from various expeditions and scribblings from my ever-present sketchbooks – all essential for this eager student on each new trip. I'll dig into my fossil collections and share some of my favourite specimens. Friends and co-explorers will join in too, contributing their images and thoughts most generously. They are an integral part of the fabric of this story – and my life.

I hope this book ....

- Puts extinction into context. Extinctions have been key in shaping life as we know it, but mass extinctions did not have a significant effect on life when considered over its full duration.

- Shows how living things thrive and change at vastly different speeds. Some, like crinoids and horseshoe crabs, barely seem to be evolving at all, whereas others – like humans – have emerged and become highly distinctive in the geological equivalent of the blink of an eye.

- Reveals the seamless continuity of life, which is obscured by traditional representations of pre-history – which must change – and puts the human animal firmly in its proper context.

- Suggests that Climate Change can be a matter of choice. Climate change is not new; it was the eventual killer in mass extinctions. Our environment changes naturally, and constantly, but can also be changed – we can adjust the dials.

We'll look beyond the media obsession with extinction and climate change to celebrate the continuity of life over time.

At the end of it all I hope you will agree that anyone who says we are living in the middle of the "Sixth Mass Extinction", doesn't fully appreciate what the other five really must have been like.

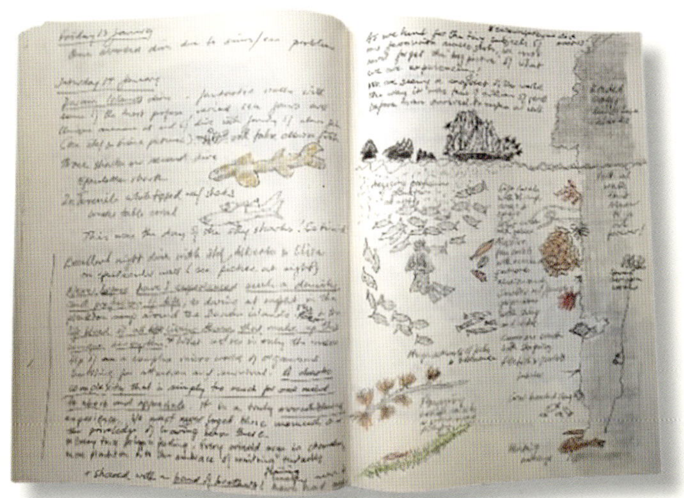

CARIBBEAN
MAD-A GAS-CAR
Bunaken Marine Park
NORTH SULAWESI
August 2004
TELUK A-YUI
WAIGEO
LAKE ST. LUCIA
FOSSIL EXPEDITION
AUGUST 2007
INDIAN OCEAN
CRYSTAL FOSSIL
... LAND!
PNG
PARADISE
SPORT
6-13 Dece
15-26 JANUARY
Sodwana
NOVEMBER
REEF!
Echidna polyzona
(Barred moray)
St. Vincent
St Vincent & the Grenadines
Union Island
Grenada
Grand Anse
Prickly Bay
Mustique
Basil's Bar
Cays
Cuttlefish
Spotted Sea hare
Aplysia dactylomela

Finally, I hope that this book leaves you with a sense of perspective about our part in life on earth, how short a time *Homo sapiens* has spent here and how arrogant we are about our own importance and sustainability. In the words of Richard Leakey "There is no law that declares the human species to be immortal".

Perhaps the story of my exploration will inspire your own – and amplify your curiosity about the surprising continuity of life. If I'm totally honest about it, this book is nothing more than an unbridled celebration of life, over time, in that eternal ocean.

To start the journey let's try to come to terms with the most misunderstood four-letter words ever – *life and time*.

While reading this book – try to keep the fold-out chart at hand. If you are unfamiliar with the flow of deep time and terms such as *Jurassic* etc, the chart will help provide a temporal context for your curiosity and exploration of the continuity of life.

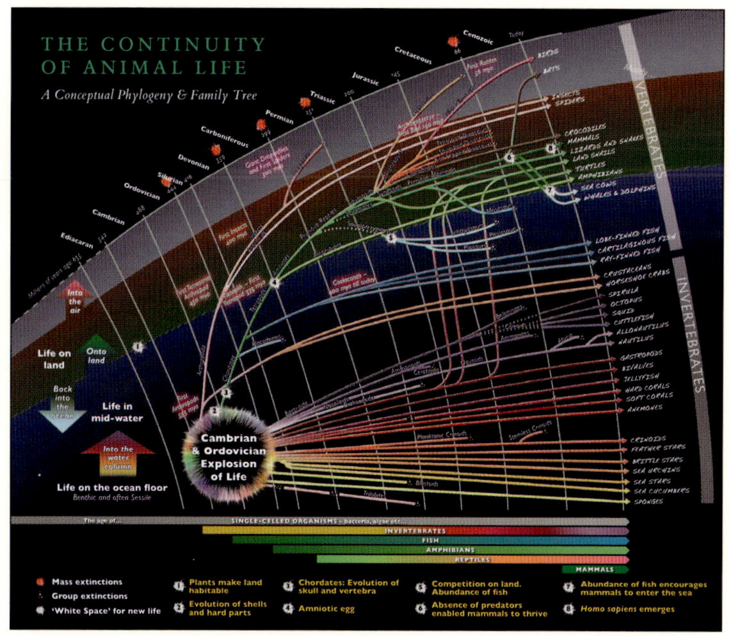

LIFE

Bacteria, energy, cells, oxygen and three domains of life

GIANT CLAMS TRAP SEA DIVERS IN GRIP OF SHELLS

Shells of huge clams found off the coast of Papua often weigh more than 400 pounds. Divers who accidentally step into

Copyright, Frank Hurley, through International

Giant Clam in Coral Reef Off New Guinea; Powerful Crushing Lips Partly Open

the open lips of the monsters are not infrequently held with such force that they cannot release themselves and are drowned. The shells close with such force that they serve as gigantic traps.

Source: Modern Mechanix, May, 1924.

Life as we know it depends on the air we breathe, specifically oxygen. Billions of years ago single-celled organisms called stromatolites made our kind of life possible by filling an atmosphere devoid of it with oxygen. Today, similar simple organisms give life and colour to a plethora of life forms in the ocean.

Giant clams of the genus *Tridacna* are common inhabitants of shallow Indo-Pacific coral reef communities and the largest shelled molluscs in the world, some larger than 2 m across. They have been around for hundreds of millions of years.

In folklore, their brilliantly colourful mantles were said to attract pearl divers and trap their arms or legs as their powerful 'jaws' closed around them. Wikipedia cites the WWII *US Navy Diving Manual* including detailed instructions on how to release oneself from the grasp of a giant clam by cutting the adductor muscles.

In remote Papua New Guinea villages, I was told stories about bivalves that swallow men whole. The giant clam shells are seen to be trappers and takers of lives, even while they are considered a spiritual source of life – the luminosity of the mantle is believed to be particularly so. The texture of the shell is believed to relate to maleness – a coagulation of semen and blood – which hardens in the woman's womb to form the hard-white bones of a foetus – analogous to life itself. My practical experience confirms the story of predatory giant clams to be the stuff of legend, but the spectacular diversity of colours and patterns in their mantles is certainly not – it is caused by symbiotic algae that grow amongst the mantle tissues.

By day, the giant clam opens up and spreads out its mantle so that the unicellular algae receive the sunlight they need to photosynthesize. Adult clams get more than two thirds of their nutrients from the algae and the rest from filter feeding.

These are similar to the bacterial stromatolites that created our oxygen-rich atmosphere billions of years ago. Stromatolites are just one of many forms of microbial structures formed on the surface of rocks or on the water surface. They are indeed one of the oldest traces of life to still thrive today, notably in Australia.

Terri with an open giant clam. Photographed in Milne Bay, Papua New Guinea. Fossil specimens from Kenya and the Philippines, around 15 million years old, can be more than 2 m across.

Two billion years ago one cell swallowed another and this symbiotic structure was the start of complex life.

Lynn Margulis

Before stromatolites

The earliest organic molecules must have required constant direct energy input from the sun, as they would have been unable to retain or store energy. Once the sun went away, even for a night, they would 'die', to be 'born' again the next morning. Nevertheless, in using solar energy to perform material conversions they were technically *alive*.

The next evolutionary step, the first 'organism', might then have been the first 'thing' that could store enough energy to keep going through the night, even at a highly reduced rate.

The advent of the first cells about 3.5 billion years ago, and of cells with a nucleus about two billion years ago, through the happenstance fusion of two unrelated microbes, signalled the start of complex life.

The cell is the basic biological unit of all present-day living organisms. It's the smallest unit of life. All organisms are composed of one or more cells and these cells contain in their nucleus the operational and hereditary information that an organism needs to live and reproduce. All cells, except the first of course, come from pre-existing cells.

Life is thus an unbroken chain of energy, a glowing baton handed on trillions of times by a runner who then slows and dies. If that chain were broken for even an instant and no energy transfer from one cell to another took place anywhere, life on earth would cease totally.

Through the ancient process of photosynthesis, the sun continues to power almost all life on earth today. Without that constant input of energy to the base of the food chain, life would fail within a few years. Well, almost all – we'll learn more about some 'alternate' life forms later.

The first cell with a nucleus was perhaps the most important and dramatic event in the history of life.

Ernst Mayr

Living stromatolites exposed at low tide, Shark Bay, Western Australia.

Oxygen: The fire of life

Life is energy in self-propagating forms.

A classical definition

Today, earth's atmosphere consists of about 21 percent pure oxygen and humans and most other animals depend on it, in the air or water, to breathe and produce energy necessary for life.

However, this level of oxygen in the air is a relatively recent development in earth's history. Until around 600 million years ago, our atmosphere was composed mainly of nitrogen and carbon dioxide – dating back to earth's formative volcanic activity more than four billion years ago – with less than five percent oxygen present.

Stromatolites represent the colonies of cyanobacteria that invented photosynthesis.

Fortunately for us there are stromatolites, one of the earliest and most prolific forms of unicellular life, that grow in shallow-water communities and are formed by layer upon layer of cyanobacteria. These oceanic cyanobacteria take in solar radiation and carbon dioxide, use the carbon for energy and expel oxygen as waste – they literally invented photosynthesis.

Initially, in the Archean world around 3.5 billion years ago, cyanobacteria seem to have been quite rare. It was a time when all life was simple, unicellular and anaerobic. Life's metabolism did not require oxygen – quite understandably, as the atmosphere was largely devoid of it.

Fossil stromatolite, *Conophyton basalticum*,
Cambrian 500 million years ago,
Antrim Basalt – Katherine – Northern Territory, Australia.

Fossil stromatolite,
Archean 3.4 billion years ago,
Strelley Pool Formation, Pilbara, Western Australia.

Fossil stromatolite, unpolished, as found in the field.
2.5 billion years ago,
Silver Hills farm, Gauteng, South Africa.

Fossil stromatolite, one of the oldest known fossils on earth.
Archean 3.49 billion years ago,
Dresser Formation, North Pole dome, Western Australia.

Under pink skies and in green seas, these simple cyanobacteria continued to pump oxygen into the atmosphere, slowly but surely. To the anaerobic life forms oxygen was pure poison but it wasn't a serious problem for them until the rate of oxygenation started to increase dramatically and atmospheric oxygen eventually stabilised at around today's levels.

This is popularly called the *Great Oxygenation Event* – not a single event at all but a period which lasted about two billion years – but it is so important that it even has its own recognised TLA (three-letter acronym): it is known as GOE in the academic literature, so I'll use that here.

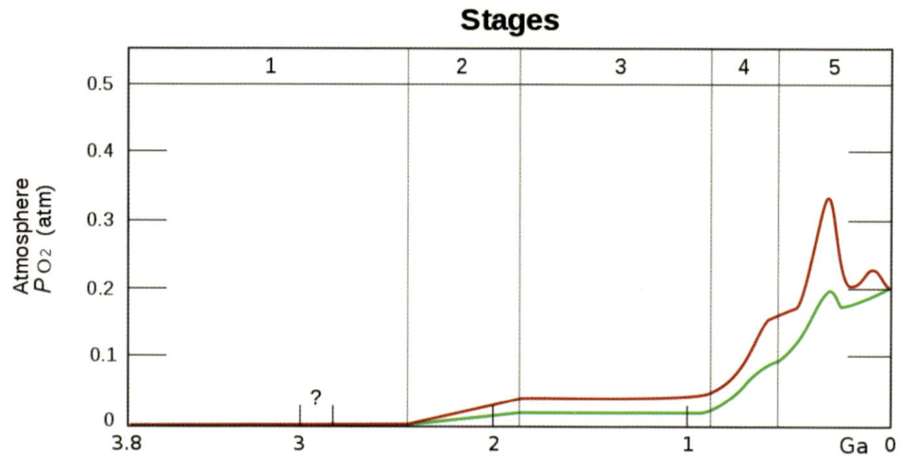

Estimated oxygen build-up in the atmosphere during the period 3.8 to 0.5 billion years ago. The upper red and lower green lines represent the range of the estimates. The stages are measured in billions of years ago.
Source: Oxygenation-atm.svg: Holland and Loudubewe.

While the GOE began around 2.4 billion years ago, cyanobacteria were present half a billion years before then, and their presence can't account for all of its effect. Before the GOE, any free oxygen that cyanobacteria produced immediately reacted with dissolved iron compounds in the water, to form insoluble 'rust', iron oxide, in thin layers on the ocean floor. This created the *Banded Ironstone Formations*, found around the world, and especially in South Africa and Australia. These include some of the oldest known rock formations and they account for more than 60% of global iron reserves. These formations emerged again from 750 – 500 million years ago.

It took cyanobacteria more than two billion years to create an oxygen-rich atmosphere suitable for the kind of life we know today.

The GOE started when these 'oxygen sinks' became saturated, at which point oxygen produced by the cyanobacteria was free to escape into the atmosphere.

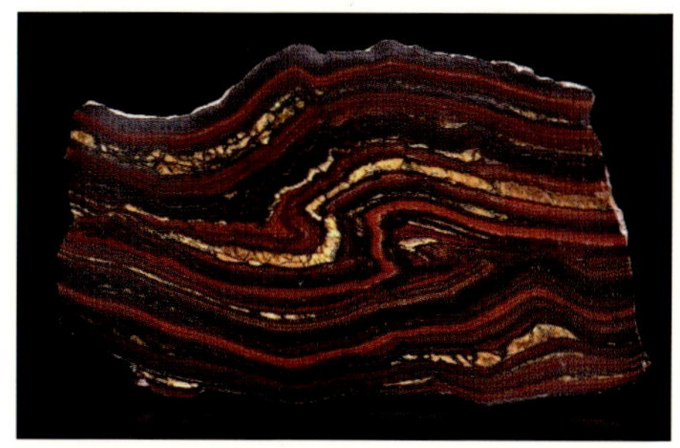

Fossil microbial sediments, 2.7 billion years ago, Ord Ranges, Pilbara, Western Australia.

During the Great Oxygenation Event, cyanobacteria and oxygen were responsible for one of the most significant mass extinctions in earth's history – although it is never mentioned as one of the top five.

The effects of the GOE and higher oxygen concentrations were vast:

- SNOWBALL EARTH: Dropping CO_2 levels caused severe glaciation that froze much of the planet. The first lasted 300 million years from 2.4 billion years ago, and further events occurred at 717–635 mya and 580 mya. They ended when volcanic eruptions created greenhouse effects that warmed the atmosphere again.

- MASS EXTINCTION: Oxygen was toxic to existing anaerobic organisms and caused their almost complete destruction. The GOE has not been called a 'catastrophy' for nothing.

- IRON + OXYGEN = RUST: The higher oxygen levels changed the nature of chemical interactions between geological substrates (rocks, clay etc), air and water – including the creation of the Banded Ironstone Formations

- NEW LIFE: The widespread availability of oxygen increased the free energy available to living organisms and new opportunities for diversification.

Alternate life forms

The deep ocean is almost devoid of life, because so little sunlight and food can be found there. Stromatolites and other microbial mats could never have developed there. Yet, the areas in the ocean around deep hydrothermal vents are bustling with strange life forms. These, and some hot mineral springs, are the only places where the source of energy for life is the inorganic earth itself.

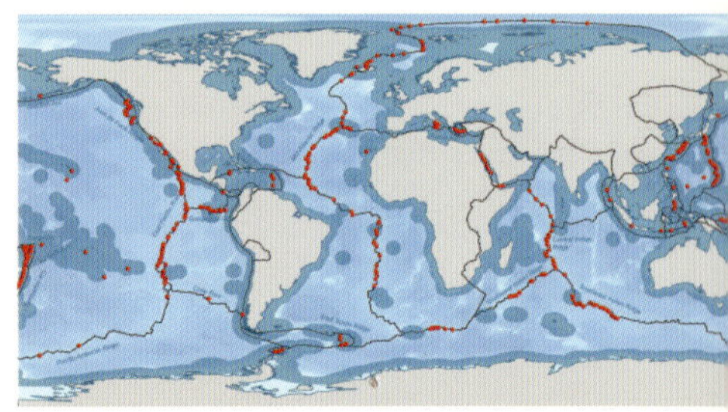

Distribution of hydrothermal vents, in red.

In these areas in the deep ocean where tectonic plates are moving apart, far from light, super-heated and mineral-rich water supplies nutrient-rich chemicals that microbes

feed on and form the base of a food chain for a diverse community of organisms. First discovered in 1977, over 300 animal species have now been identified as living around them. Almost all of them are unique to the vent environments and were previously unknown.

The large tube worms living there absorb chemicals from vent fluids to feed bacteria living in them. In return, the bacteria provide the carbon necessary for the tube worms to live. The organisms that thrive there could not survive in the oxygen-rich world above them, although functionally their lifestyle is similar to the algae in the mantles of giant clams.

Some of the heat-loving microbes that inhabit these vents are amongst the most primitive organisms known on earth, classified in the domain Archaea, the *third domain of life*, with which most of us are least familiar. These single-celled organisms have no cell nucleus. They are as different from bacteria as bacteria are from all other organisms.

A large concentrations of the tube worm Riftia pachyptila, with anemones and mussels, on a hydrothermal vent in the Galapagos Rift. From the 2011 NOAA Galapagos Rift Expedition.

For these strange life forms, the source of energy is the inorganic earth itself.

Archaea were initially thought to live only in extreme, harsh environments, such as hot springs and salt lakes. Since then they have been found to be common in the soil, the ocean and on us, in our mouths, on our skin and essential to our digestive system. These simple cells without a nucleus, live alongside complex organic molecules. It is thought by some that life could have begun at such deep-sea vents; certainly, the vent communities are ancient and have changed little over hundreds of millions of years.

It is worth taking a little detour here to reflect on the three domains into which all life is classified – *Archaea*, *Bacteria* and *Eukaryotes*.

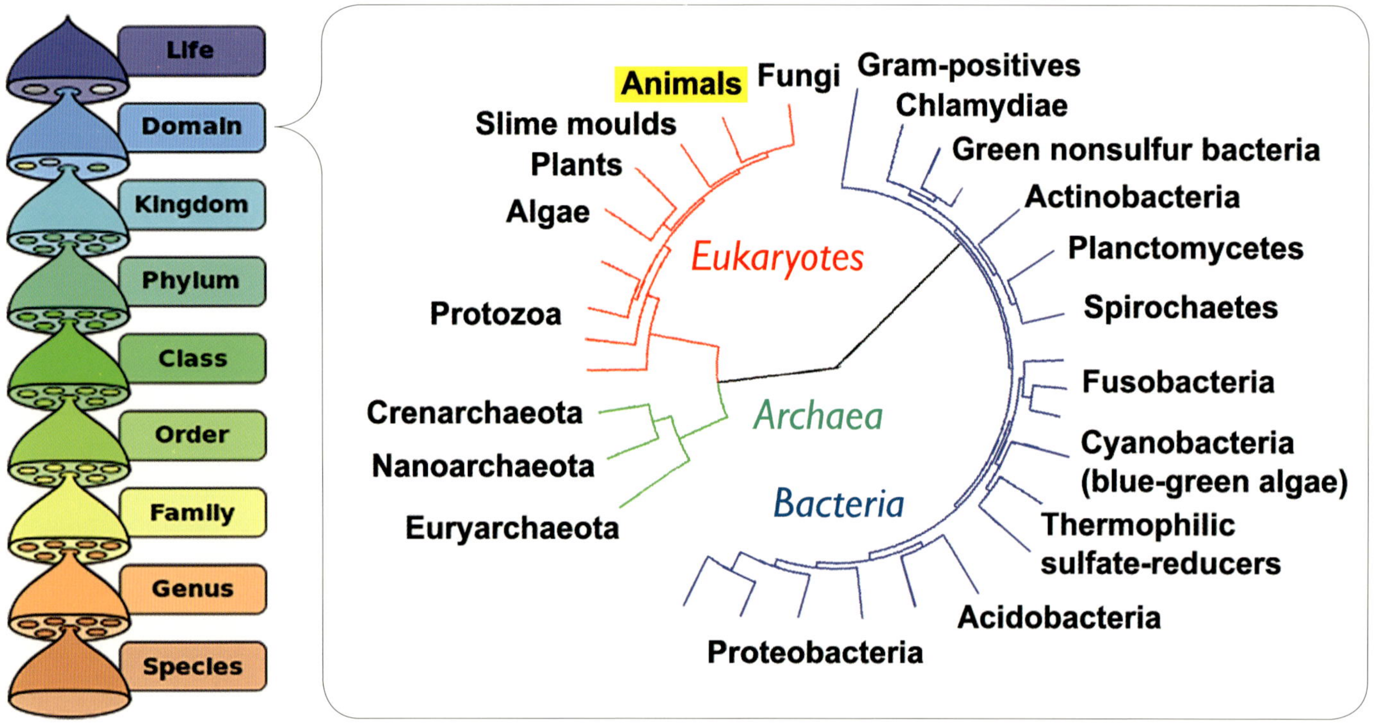

Phylogenetic tree of the Three Domains of Life showing the relationship between the eukaryotes and other forms of life. A highly resolved Tree of Life, based on completely sequenced genomes.

From this chart it is evident that this book only focuses on a very small part of the diversity of life – one tiny spoke in the cladistic diagram above – the *Animals*, a part of the domain *Eukaryotes* that also includes *Plants* and those 'more basic' inhabitants of the world around us – the *Fungi, Slime Moulds* and the single-celled organism of the kingdom Protista, the *Algae* and *Protozoa*. This is a perspective on the diversity of life that truly puts *Animals* – and *Homo sapiens* – into their correct, and somewhat insignificant, perspective.

You might ask, why have I ignored *Plants* in this book about life in the ocean? Well, that's because plants are primarily of the land – they are extremely rare in the ocean. Many marine life forms that look like plants, such as corals, anemones and crinoids, are actually animals – and animals are what I have chosen to focus on.

The only real plants in the ocean are seagrasses, so called because their leaves superficially resemble terrestrial grasses, and which grow in shallow marine, fully saline environments. They are actually flowering plants, angiosperms, which evolved only in the Cretaceous Period, and they help provide an ideal ecosystem for many small marine species.

Seagrass beds.

Even seaweeds are not plants. They are algae. Floating Sargassum seaweed is phytoplankton, microscopic algae, and sessile seaweeds, often to be found washed-up on beaches, are in fact red, brown or green macroalgae.

Kelp is also a form of marine algae that first appeared in the Miocene, between 23 and 5 million years ago. Kelp are not plants and calling aggregations of kelp 'underwater forests' is just a good example of a proverbial red herring (and that is definitely a mixed metaphor). At best, all these algae can be called *plant-like organisms*.

A kelp 'forest' rising above a ledge with sea anemones, encrusting sponges and sea stars. Photographed off Cape Town, South Africa.

The biomass measure of life

Biomass is another measure that puts the animal world into vivid perspective. How this is measured and how accurate the figures are, or indeed can be, is always up for debate. One can measure dry mass or wet mass, which can be vastly different. Several estimates put the total live biomass on earth at between 500 and 1000 gigatons of carbon (Gt C), and here I have shown an approximation of several studies in the chart at right. You can see just how small the total biomass of animals actually is and how insignificant plants are in the ocean. Also seen clearly is the massive contribution of the 'invisible' *Archae* and *Bacteria*.

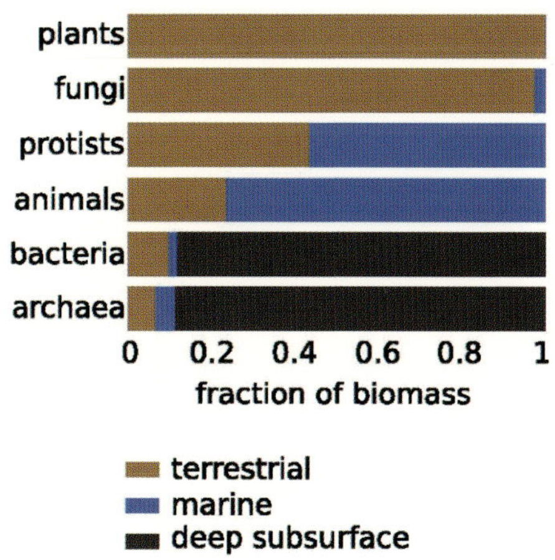

Source: www.pnas.org/content/115/25/6506

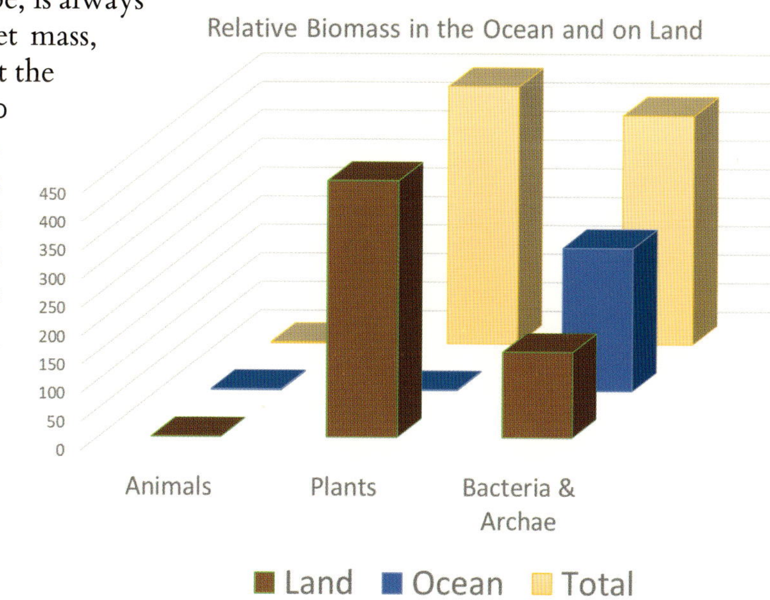

Bacteria: The ultimate form of life?

Bacteria share information. Any bacteria can take pieces of genetic coding from any other. All bacteria swim in a single gene pool. It's rather as if a human could go to an insect to get the necessary genetic coding to sprout wings or walk on ceilings.

Bill Bryson

The simple life forms that dominated the earth before the Cambrian explosion of life 500 million years ago still make up most of life on earth today – according to many estimates, around half of the total biomass.

Author Bill Bryson captured that fact most eloquently in his best-selling book *A Short History of Nearly Everything*. A chapter entitled *Small World* is in my view one of the most compelling ever written on natural science. He relates that the human body is composed of around 10 quadrillion cells and it is inhabited by ten times as many, 100 quadrillion, bacterial cells. Also, if you are an average healthy human, your skin alone is covered by more than one trillion bacteria. That works out at 100,000 bacteria in each square centimetre. He concluded that bacteria likely regard you as 'the ultimate buffet'. We humans have a history of less than one million years on this planet and we live in a pulsating ether of incredibly resilient bacteria.

Christian de Duve, who won the Nobel Prize in Physiology for Medicine 1974, put it into humbling perspective: "Given an adequate supply of nutrients, a single bacterial cell can generate 280 trillion individuals in a single day". In the same period, a mammal cell manages just a single division. And, once alive, they seem to be able to survive for ever. Microbes more than 250 million years old have already been resuscitated – older than the continents that we regard as old today.

We are just visiting their planet.

It is safe to expect that bacteria will still be here when human life has been extinct for a hundred million years. Bacteria will surely prove to be the ultimate continuity of Life.

Bacteria are made up of single, complete living cells but lack a nucleus, mitochondria and chloroplasts. They are the simplest kinds of living organisms, they reproduce by cell division and can be killed and die if nutrients cannot pass through the cell wall, or waste products cannot escape. They are considered alive but viruses are not – they have been eloquently called "just a fragment of DNA or RNA wrapped in a protective protein coat" – even simpler than bacteria. They do not consume food, have no ongoing metabolic processes, cannot reproduce on their own and remain inert, until they encounter desirable host cells. Viruses literally hijack cells and turn them into nano-machines to produce more virus particles. Without living host cells, viruses are 'nothing' – and that is why they are excluded from the three domains of life.

So, what exactly is life?

Perhaps now that we are almost at the end of our chapter on *life* we should try to define exactly what we mean by that awkward four-letter word.

It took billions of years for the earth to get ready for life. Now, after another 500 million years in the evolution of complex life, it's all around us. But what do we actually mean when we say something is *alive*? What is the set of characteristics that a 'thing' has to have in order to be considered to be so?

Widely accepted today, the five key characteristics of life are:

Biologists have failed to devise a single, universally accepted definition of life.

Robert M. Hazen 2012 *The Story of Earth*

life

/lʌɪf/

noun

1. the condition that distinguishes animals and plants from inorganic matter, including the capacity for growth, reproduction, functional activity, and continual change preceding death.
   "the origins of life"
   synonyms: existence, being, living, animation, aliveness, animateness
2. the existence of an individual human being or animal.
   "a disaster that claimed the lives of 266 people"
   synonyms: person, human being, individual, mortal, soul, creature
   "more than 1,500 lives were lost in the accident"

1. *Capacity for growth*
2. *Reproduction*
3. *Functional activity*
4. *Continual change*
5. *Death*

Richard Dawkins once suggested that if you want to understand life, "don't think about vibrant, throbbing gels, and oozes, think about information". If you apply our 21st century technological perspectives to it, then you might well say that *information/data plus algorithms plus an irrepressible desire to reproduce* may be the best definition of life.

So, the next question may be obvious... by these characteristics could computers and software be considered alive? We'd best leave that debate for another occasion.

Life is a predicament which precedes death.

Henry James

The totality of life, known as the biosphere to scientists and creation to theologians, is a membrane of organisms wrapped around Earth so thin it cannot be seen edgewise from space, yet is so internally complex that most species composing it remain undiscovered.

The opening sentence in
Edward O Wilson's 2002 book
The Future of Life

The unparalleled ability of water to dissolve and transport other chemicals makes it an ideal medium for the origins and evolution of life. All life on earth, and perhaps all life in the cosmos, depends on water.

Robert M. Hazen

Life: An understanding forever incomplete

Things are alive when we agree they are.

Paul Callomon

One thing is for sure: no matter how hard we try, the definition of *life* will forever remain incomplete. Biologically it neither includes nor excludes viruses and does not deal with pre-cellular or other non-cellular life very well at all. It does not comfortably embrace future, or as yet unknown forms of life.

Intuitively, we may realise that we humans are a recent iteration and that our experience is just a tiny blip in life on this planet, but we struggle to comprehend just how small a blip it may be. Many now-extinct life forms, such as ammonites as well as the common sea-urchin, survived thousands of times longer than humankind has so far.

Another question asked about natural life is why it continues. The sole end-game of all life forms seems to be to pass on life's energy and thus assure the continuation of life itself, but there is no ultimate logic to it. The rest of the universe (so far as we know) seems to consist entirely of systems that adhere to the laws of physics and chemistry; why does life persist in defying this, and why does it arise and diversify so readily?

Why does life propagate itself in the face of otherwise almost universal entropy?

We are only at the beginning of our understanding of life in the universe. Neither science nor religion has an answer, but so far life seems to be the exception to the universal standard, not its basis.

If we assume the goal of life to be its own continuation, and life to have begun with cells, then there is yet no explanation for the logic that led to their organization. We can't work backwards from present-day sophistication to assign inevitability to its beginning. Life is not easy to see with a telescope.

The answer to 'What is life?" largely depends on *when* you ask the question. What is certain is that the answer you give today will be different to the answer the identical question will elicit fifty years from now. And that response may well come from an intelligent machine.

What is life? The answer depends on when you ask the question.

Fifty years – that's a long time, isn't it? Just think how full your reservoir of nostalgia will be by then. Well, perhaps geologically speaking, that isn't such a long time after all.

Read on.

TIME

Clocks, quanta & chrononyms

Time is the indefinite continued progress of existence and events that occur in apparently irreversible succession from the past, through the present to the future.

A dictionary definition

It's 11:35. Before my lunch appointment with Daisy at Le Train Bleu I'll just have time to pop into the florist, pick up some flowers and then take a taxi to the restaurant. Forty minutes should do it if the Paris traffic is not too bad.

In our day-to-day lives where would we be without the organising structure that time provides? How could we manage personal and business relationships without the shared norms of minutes, hours and days? We have deadlines, targets and appointments tied to specific dates and times. It's the same if you work in a call-centre, are a legal professional, an Uber driver or a top executive. At home there are bills to pay and kids to educate and entertain. Without the clockwork diary our lives might seem 'chaotic' and 'out-of-control'.

The entries in our diaries are the cogs that drive the mechanism of life itself. The diary has become the critical organising principle of our busy lives. For the really organised amongst us our schedule for the year is filled-in in some detail on day one. Some corporate warriors complain of their diary being full of 'routine bureaucracy' on day one of the new year – "Never enough time left to do the really important stuff". A year seems to have become a convenient block of time to 'manage' and for many it is managed down to 15-minute blocks. Your doctor would normally plan to see you for no longer. If an appointment with your lawyer takes longer than thirty minutes, both parties start thinking seriously of the invoice that will duly appear.

Time is nature's way of preventing everything from happening at once.

Traditional

Many of us complain that we are slaves to time, and that there's never enough of it.

Our high-tech world has forced us to think of time in ever smaller chunks. Competition on the New York Stock Exchange's NASDAQ has been characterised as a race to reduce trading speeds to zero. Everyone wants to be on the *fastest trading platform* – and that means times measured in nanoseconds. Some professional market traders are even talking about stocks being traded in picoseconds – trillionths of a second.

Currently, the international unit of time, the second, is defined by measuring the electronic transition frequency of caesium atoms.

It is clear that without time, a lot of what we experience would make no sense at all: the speed of a train, acceleration, the apparent motion of the sun across the sky, the phases of the moon, the swing of a pendulum, and the beat of a heart.

Time is clearly also of significant personal, social and economic value ("time is money"). We are acutely aware of the finite time in each day. Activities such as business, sport, science and the arts all incorporate some notion of time into their measuring systems.

Every target requires a scale – no matter how small or large. Every story needs a timeline. Novels typically focus on moments in time, special memories or perhaps a lifetime. History stretches over multiple generations and many lifetimes.

This book covers an unimaginably large chunk of time – more than 500 million years – the extent of complex life on earth – a span impossible to comprehend when we are used to managing our lives down to minutes.

Molecular clocks

In trying to understand the evolution of species over time, and the evolutionary branches in the tree of life, scientists use the concept of a *molecular clock*, using the mutation rate of biological molecules to deduce the time when two or more life forms diverged. It turns out that DNA is a kind of clock, gradually changing and keeping time of the genetic changes.

Instead of measuring seconds, minutes and hours, molecular clocks measure the number of changes, or mutations, which accumulate in the gene sequences of different species over time.

Trillions of different clocks

The processes of life run at different speeds, both within and between organisms. The growth of your skeleton, for example, is timed differently to the pace of your breathing. A dog's life goes through the same stages of infancy, adolescence, adulthood and senescence as it's master's, but in less than a quarter of the time. For cats it's somewhat less and hamsters last only two to three years.

Think of life in the context of time as trillions of clocks, running concurrently, all running at different speeds.

Many animals are on a far shorter schedule, however. Squids rarely live more than a year, and their more intelligent relatives, the octopuses, don't fare much better. Insects have the briefest lives of all; though their eggs and larvae may live for a few years, some species have adult lives measured in days or even hours.

Although a week comprises all of the lifespan of many insects, it must on the other hand seem like a brief moment to the Aldabra Giant Tortoise, whose lifetime can exceed 150 years, or an Icelandic deep-sea clam whose quiet existence can span more than five centuries. Each is nevertheless an iteration of life that involves at least one passing of the baton to the next generation and usually many.

Chrononyms are made of this

Chrononym – noun
A term for a specific period of time.

Dictionary definition

To make time more manageable we tend to give names to various time segments – say *Monday* or *Tuesday*. "I'll see you at the restaurant on 01.12.18.12:00" is so much less helpful than "at noon next Tuesday".

Even a year is often too long a period for the human mind to get to grips with, so we invent terms to break it up into more meaningful slices such as: 'spring', 'summer', 'autumn' and 'winter', terms that add specific physical properties to the elusive concept of time. "We'll see you next summer" suddenly conjures up possibilities more attractive than promising to meet again in 180 days.

Now, at the start of my eighth decade on the planet, I find others referring to the 'winter' of my life and I hastily try to convince myself that this actually feels more like 'autumn' and that there is much more to look forward to.

These convenient *chrononyms* simplify our communication, memory and recall. Think of the vast numbers of these that we use: a *lifetime*, a *generation* – quite imprecise words but they give us an immediate sense of the time we are talking about.

And, so it is with *geological time*. Here we are dealing with truly incomprehensible numbers. Hundreds of millions of years are beyond what we can imagine. The concepts roll off geologists' tongues as if they can be understood.

To make geological time easier to comprehend we also turn to chrononyms to group chunks of time, such as the *Jurassic* and *Cretaceous* Periods, convenient chunks of roughly 50 and 80 million years. Many of the named Geological Periods are conveniently separated by major extinctions.

But these are still too 'chunky'. To create more granularity and definition, scientists have taken a closer look at the rock strata in which fossils were laid down and have given them names – many of which hint at where these were first identified – *Devonian* for those strata found in Devon, England, and *Maastrichtian* after fossil discoveries made in Maastricht, Holland.

Does time exist?

When one can identify a series of cyclical events, such as the movement of a pendulum, happening in a standard unit of time such as a second, this is highly useful in the conduct of sophisticated experiments as well as in everyday life. Such practical uses of *time* leave aside the question whether there is actually something called *time*, that flows apart from the clocks and the actual counting performed, and that can be measured. Time is convenient, even vital, but we are still not sure exactly what it is.

As seen through the window of *clocks*, time is an arbitrary system of measurement that we invented. It's perhaps not surprising that in hard physics, time is now unambiguously defined as 'what a clock reads'. But it hasn't always been like that.

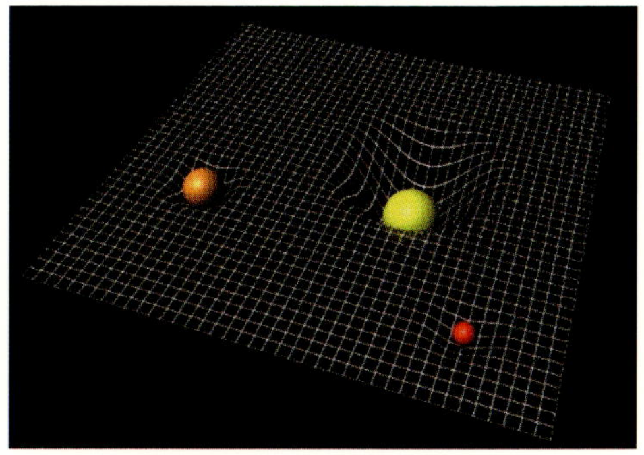

Conceptual representation of Einstein's Space-time curvature.

Newton's work influenced the 18th century view that time is part of the fundamental structure of the universe – a dimension independent of events, which occur in sequence. After Albert Einstein published his *Theory of Relativity* early in the 20th century, *Newtonian Time* was superseded by the notions of time as a dimension of space-time, and of dynamically curved four-dimensional space-time

Recently, an opposing view emerged that time does not refer to any kind of 'container' that events and objects 'move through', nor to any entity that 'flows'. This view is based on the science of *Quantum Gravity*, in which *space-time* and *gravity* are considered to be *emergent phenomena*, and that time is neither an *event* nor a *thing*, and thus is not itself measurable.

Emergent properties arise due to the interactions among the parts of complex systems. These properties are not contained in any of its parts. Emergence plays a central role in the behaviour of complex systems such as the weather.

"I think we now understand that space-time really is just a geometrical representation of the entanglement structure of these underlying quantum systems," Mark Van Raamsdonk, a theoretical physicist at the University of British Columbia has been quoted as saying, somewhat obscurely. Did you get that?

As an aside, the definitions emerging from Quantum Gravity seem to make 'time travel' impossible. That conclusion may shatter many a dream. What a pity. Perhaps a future discovery will help rekindle our fertile imaginations in this regard.

Ancient worlds: Deep time

Deep time is so alien that we can really only comprehend it as a metaphor.

Stephen Jay Gould

In most people's minds, values beyond those that can be related to a human scale are abstract. A million years has no emotional connection; if you sentence someone to a hundred years in prison instead of five, you might as well make it a million. Both are unimaginable periods. Contrast that, however, with being sentenced to ten years in prison instead of one.

We can briefly trick the mind into glimpsing vast lengths of time using visual and verbal analogies, but they only work with numbers and terms that people can emotionally relate to. To give some idea of how big a million is, you could start counting seconds. A million seconds will take you eleven and a half days. A billion seconds will take 31.7 years, but most of us can't begin to imagine how long that feels.

Stephen Hawking confirmed that "all the evidence seems to indicate, that the universe has not existed forever, but that it had a beginning, about 15 billion years ago. This is probably the most remarkable discovery of modern cosmology. Yet it is now taken for granted. We are not yet certain whether the universe will have an end".

Few minds can sustain visions of such scales, such as the death of the Sun in several billion years, because we have no emotional investment in their ending. Stephen Jay Gould once said that he had taught the discovery of time for twenty years and that deep time was so difficult to comprehend that it was a major stumbling block to our understanding of life on earth.

To give some idea of how big a million is, you could start counting seconds. A million seconds will take you eleven and a half days. A billion seconds will take 31.7 years, but most of us can't begin to imagine how long that feels.

Paul Callomon

In 240 BC Eratosthenes was able to calculate the circumference of the earth with surprising accuracy – 39 690 km vs the actual 40 075 km – accurate to 1%. Almost 2000 years later, in 1650, Archbishop Usher, calculated that the first day of creation fell on 23 October 4004 BC – making the age of the earth around 6,000 years. That 'fact' was readily accepted by the church and is still considered as fact by some today, even though it is out by billions of years.

Time flies like an arrow. Fruit flies like a banana.

An anonymous saying that is often given to students of the English language to illustrate syntactic ambiguity, and in word play as an example of punning, double entendre, and antanaclasis.

The King of Time is dead

It's 6am on Wednesday 14 March 2018 and the news informs me that "the greatest living scientist is dead". Stephen Hawking, that celebrated explorer of black holes who created extraordinary bridges between the academic and popular cultures, is no more.

His best-selling book *A brief history of time*, rated as the most popular scientific book of all time, sold more than 10 million copies. It also has the unenviable award for the least read book ever. Clearly it wasn't quite brief enough for most readers. So it appears that the most authoritative book on time could not even raise enough enthusiasm for people to really understand its subject.

One of science's great contributions to humanity is the ability to suppress the *self* from our world-view, to enable us to step back from our reality and see it from the outside in. It does not seem to make the task of understanding *time* and *life* any easier.

With quantum physics having jettisoned it as a fundamental structure, perhaps *time* has turned out to be much simpler than we thought. We humans are hugely invested in *time* to enable us to achieve what we want – even if it does not empirically exist.

Just think of *life* over *time* as trillions of clocks (in millions of different life forms) each running at their own pace – and YOU are at the centre of that maelstrom of experience – even if that is a somewhat anthropocentric perspective.

Understanding the interaction between the two four-letter words *life* and *time* is where the next chapter takes us.

Exploring the nature of time leads us to understand something about ourselves.

Carlo Rovelli

ENLIGHTENMENT

Making sense of life over time

The age of the universe is around 15 billion years and the earth is estimated to be around 4.5 billion years old. The oceans were formed 100 million years later, and first simple life emerged around 3.8 billion years ago. For the first three billion years all life forms were aquatic. Time spans like these are nowadays accepted as fact, yet it is easy to forget that less than 200 years ago much conventional thinking was that the earth, and everything on it, was no older than 6000 years.

According to Genesis *"In the Beginning God Created the Heaven and the Earth"*. This unambiguous statement prompted many attempts at calculating an actual and specific date. Seventeenth century academics pursued this goal fervently and both Kepler and Newton came up with estimates of less than 10 000 years. In the end *The Chronological System of Archbishop Usher* emerged as the most popular. The Irish theologian placed the creation of the earth in the year 4004 BC – on the 23rd of October to be exact. He relied largely on Jewish tradition to fix the date. In his chronology he dated *"The Deluge"* or *"Flood"* at 2348 BC.

The Bible was not alone in establishing the idea that in ancient times much land was once under water. Wikipedia lists a range of sources from hundreds of years BC, including Xenophanes, Herodotus, Eratosthenes and Strabo, who all held this view based on fossils they had seen of marine organisms. During the Middle Ages, fossils were discussed by Persian and European naturalists, and the Chinese naturalist Shen Kuo proposed a theory of climate change based on evidence from petrified bamboo.

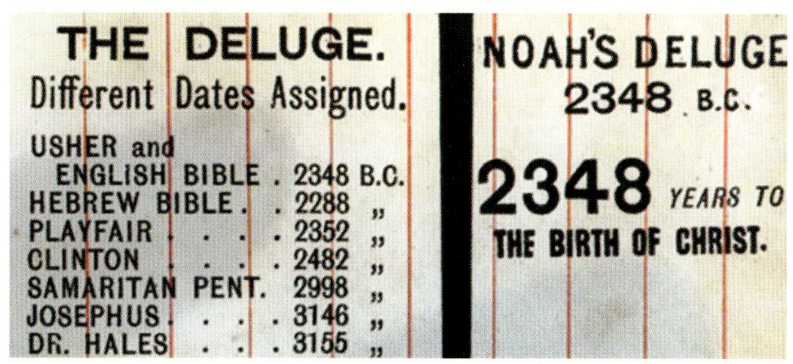

This engraving by an unknown artist first appeared in Camille Flammarion's *L'atmosphère: météorologie populaire* 1888. The associated caption translates as 'A medieval missionary tells that he has found the point where heaven and Earth meet'.

Curiosity – entirely human

Humans evolved only very recently in geological time, but our level of reflectivity and self-awareness is different by an order of magnitude to that of other animals. Curiosity is one of the defining characteristics of our species. We have always searched for a complete understanding of our environment and the context within which we exist – our personal 'universe'.

In his 2016 book *The Return of Curiosity* Nicholas Thomas describes curiosity as "an unstable attitude" marked by an eagerness to encounter what is new and unfamiliar, an openness to difference and a willingness to suspend judgement. But, he says, curiosity has not always been seen as a positive character trait.

Edmund Burke's classic 1757 treatise *A Philosophical Enquiry into the Origin of Our Ideas of the Sublime and Beautiful* dismissed curiosity as the "skittish attitude of children" and "arousal by mere novelty" and mirrored it against the apparently more valid "mature and masculine exercise of reason and judgement". It never seemed to enter Burke's head that both may be possible, and even essential, for real creativity. That this may not be an either/or question at all.

Thomas also cites the *Curious George* stories, first published in France in 1939, as an example that teaches children that "curiosity gets you into trouble". In the first of the *Curious George* stories, George, an orphaned monkey, immediately demonstrates this lesson. A man in a large yellow hat spots him in the jungle through his binoculars and decides he would like to take the monkey home with him. He puts his hat on the ground and hides behind a tree. George, ever curious, comes down from the tree to look at the large yellow hat. He puts on the hat, but it is so large that he can't see. That gives the Man in the Yellow Hat the opportunity to capture George and put him in a bag. You see how bad curiosity is for you.

The oceans and cultures: Gods, spirits and the unknown

The cradle of humankind was also the cauldron of curiosity. So much to wonder about, so much inexplicable. Where does the sun go at night? How is it reborn in the morning? Contemplating such questions alone is difficult enough. When a community ponders such questions in the absence of any obvious facts, myths are born.

Perforated *Nassarius kraussianus* shells discovered at Blombos Cave, South Africa, believed to be about 70,000 years old and one of the first examples of personal adornment.

Most early communities were established around coasts, lakes and rivers – water facilitated transport, and marine life provided essential nourishment, while shells were a key element in development of human culture.

At Blombos Cave on South Africa's south coast a cluster of 24 perforated *Nassarius kraussianus* shells dating back 70,000 years were recently uncovered. It is believed that they were deliberately pierced through the aperture, probably with a bone tool. Analysis of wear patterns on the shells show that the shells were strung, perhaps on cord or sinew, and worn as personal ornament. Ochre has been detected inside some of the shell beads, indicating the use of a colouring agent. This is one of the first examples of self-awareness and personal adornment and was likely an important factor in cognitive evolution.

The Etruscan civilisation that lived in Italy between 700 and 400 BC, decorated their dishes with fish and cuttlefish.

Until recently, the first use of personal ornaments was thought to have occured with the arrival of *Homo sapiens* in Europe, some 40,000 years ago. Most academics now believe that marine shells were used as beads across Africa at least 30,000 years earlier than their first use in Europe.

Portrait of Andrea Doria as Neptune.

To early people rationalising the meaning and rhythms of the world around them, the oceans were an important element, strange and unknown, feared and revered. A dystopian sea demanded gods on whom you could call for protection against the demons, devils and dragons that might threaten you, your families and your endeavours.

The powerful water deities of Greek and Roman cultures are well known to Western students. Neptune, the Roman god of freshwater and the sea was the counterpart of the Greek god Poseidon. In the Greek-influenced Roman tradition, Neptune was the brother of Jupiter and Pluto; the brothers presided over the realms of Heaven, the earthly world, and the Underworld.

Greek mythology was characterised by a vast plethora of gods related to water, the sea and oceans. Poseidon was the god of the ocean, king of all sea gods, and he controlled every aspect of the seas. Amongst his other responsibilities he was also god of rivers, storms, flood and drought. Oceanus, one of his under-gods, was god of the earth-encircling river *Okeanos*, said to be the font of all the earth's fresh-water. Tethys was goddess of the sources of fresh-water, and the mother of the rivers, springs, streams, fountains and clouds. Thalassa, another rather more obscure deity, was the primordial goddess of the sea.

Poseidon, from a 2600-year-old mural in Agrigento.

Carved image of Scylla, in the Naples Archaeology Museum.

Legends and mythologies abounded in ancient Greece to support the roles of these myriad gods. In Homer's *Odyssey*, sailors were told to be aware of terrible threats when they passed through narrow channels. On one side lived Scylla, a multi- headed monster with bad attitude. Homer described this cave-dwelling beast as possessing twelve legs and six necks, each with a ferocious man-eating head that sported three rows of teeth. Opposite Scylla, on the opposite side of the channel, lived her counterpart Charybdis. The two sides of the strait were within an arrow's range of each other – so close that sailors attempting to avoid Charybdis would pass dangerously close to Scylla and vice versa.

Dystopian indeed.

On this page are several illustrations from the original 1870 edition of *20,000 Leagues Under the Sea* by Jules Verne.

Dystopian seas: 18–19th century mythologies

Over the ages and in popular culture, the Scylla of Greek legend seems to have evolved into *The Kraken*, a legendary sea monster of giant size, which was first described by the Dane Erik Pontoppidan in the middle of the 18th century. It was said to dwell off the coasts of Norway and Greenland.

Authors over the years have postulated that the legend originated from sightings of giant squid that may grow to 12–15 meters in length. The sheer size and fearsome appearance attributed to the Kraken have made it a common ocean-dwelling monster in many fictional works ever since.

It has also been suggested, however, that both Scylla and The Kraken could have been mere extrapolations of encounters with ordinary octopus or squid.

On this fertile ground was built some of the most spectacular of Jules Verne's novels. His 1871 book *20,000 Leagues Under the Sea* captured readers with a generous plethora of fabulous dystopian art by Édouard Riou and Alphonse de Neuville, photos below right. Looking at them with what we know now, their 19th century illustrations could well have been images of life in a Jurassic ocean, except that the octopus would have had a shell – ammonites were prolific in Jurassic and Cretaceous seas.

Colossal octopus by Pierre Denys de Montfort 1801.

Dystopian seas : primal cultures

Fabulous stories, myths and legends surrounding the ocean were not limited to European cultures; they seemed to be pervasive almost everywhere.

In ancient Asian mythology, starting from around the 7th century, *The Tide Jewels* were magical gems that the *Sea God* used to control the tides. These myths interconnect with dragons and sea-monsters in Indonesian mythology, the *Wish-Fulfilling Jewel* in Japanese lore, the magic jewels of Nāga kings in Hindu mythology, and Chinese dragon mythology.

This spectacular woodblock print by Utagawa Kuniyoshi, from the Edo Period in the 17th-19th century, depicts Tamatori, who has penetrated the Dragon King's palace to steal the jewel, on his escape plunging through waves with outstretched knife, pursued by a dragon and an assortment of irate marine life.

In Melanesian primal cultures and other Pacific island groups, there were many water gods, spirits and dreadful monsters lurking in the sea. Cautionary mythologies were associated with them, at least until Western missionaries outlawed such pagan beliefs in the 18th and 19th centuries.

Before that, time was never measured by clocks and these cultures likely saw the commoditisation of time as somewhat subversive and counter to the stability of their communities.

I have some examples of these primal belief systems in my personal collections of Melanesian art. The story of the *Adaro*, a malevolent merman-like sea spirit from the mythology of the Solomon Islands, north-east of Australia, in the Pacific Ocean is one of my favourites.

Adaro is said to arise from the wicked part of a person's spirit, which is divided between the *Aunga* (good) which dies, and the *Adaro* (evil) which stays on as a ghost. An *Adaro* is variously described as a man with gills behind his ears, tail fins for feet, a horn like a shark's dorsal fin, and a swordfish- or sawfish-like spear growing out of his head. He may also travel in waterspouts and along rainbows and is said to kill unwary fishermen by firing flying fish at them.

The spiritual artworks and functional objects made to represent the *Adaro* clearly fire the creative juices. So pervasive is this myth that the *Adaro* even made it onto the Solomon Islands stamps and coins.

Anything that can be imagined is real.

Pablo Picasso

Changing perceptions

As early civilisations drew on supernatural and divine forces to explain origins and outcomes that could not be accounted for by alchemists or physicians, they were always careful to portray themselves as subordinate to the divinities, spirits and gods of their particular culture. Certainly, there was a consistent view that the physical world was created by some superior force at some point in time and would one day end.

This view was increasingly eroded by the ability of humans to understand and predict phenomena around them more rationally, based on a growing body of transmittable knowledge, especially printed books. There emerged a constant tension between *beliefs* and *facts*.

Onto this stage entered the study of fossils – eventually deduced to be the remains of ancient beings – plus the emergent sciences of geology and palaeontology.

Man's mind, once stretched by a new idea, never returns to its original dimension.

Oliver Wendell Holmes

The impact of fossils

Fossilisation is a process where minerals in the soil or water replace buried bones and shells of animals, over thousands and millions of years, after the soft parts have typically been eaten by predators or bacteria. When these remains are exposed millions of years later, we call them fossils – evidence of previous ancient life.

This happens best when dead animals are covered in fine sediment that excludes oxygen, making the sedimentary deposits on the sea bed one of the most prolific sources for fossils. Bones and shells on land and in fresh water tend to dissolve rather than become entombed, so relatively less is known about their historical abundance and diversity. This is especially so with mammals, reptiles and other terrestrial animals, whose fossils are all rare. Whole animals are only preserved on land where rapid burying conditions like those on the sea bed occur, such as in a desert dust storm or a volcanic eruption.

For this reason, we understand the ancient history of the marine animals with shells far better than that of any other animals.

Fossil:
1. the remains or impression of a prehistoric plant or animal embedded in rock and preserved in petrified form.
2. a person or thing that is outdated or resistant to change.

Dictionary definition

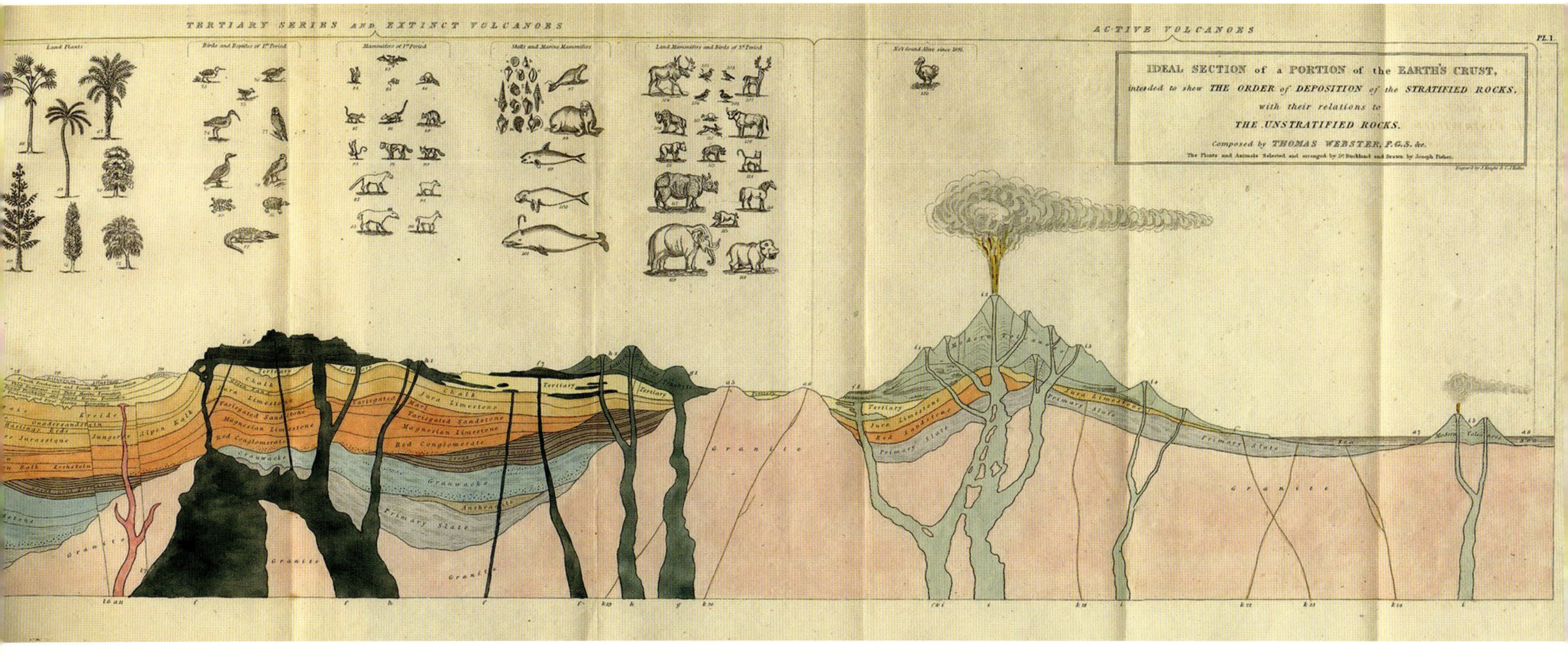

Synoptic View of the leading Phenomena of Geology: William Buckland, *Geology & Mineralogy Considered with Reference to Natural Theology*. Vol. 2. William Pickering London 1836.

It is unfortunate that only the hard parts of animals tend to fossilise. It is extremely rare to find fossils or even imprint evidence of soft-bodied animals such as jellyfish – although they do occasionally occur. If, for example, a 20-million-year period was dominated by huge swarms of jellyfish, there would most likely be no fossil evidence of them ever having existed.

During the 19th century it became popular pastime to look under one's feet for *magic stones*. But, how to interpret what these meant?

The faithful vs the rational

Faith and rationality are two ideologies that exist in varying degrees of conflict or compatibility. Rationality is based on reason and facts. Faith is belief based on inspiration, revelation and constructed authority. Faith includes beliefs that are held in spite of, or alongside, reason or evidence.

Most peoples' thinking is somewhere along a scale between the two extremes. Where are you?

While some may hold extremely polarised views, in reality, both extremes are unlikely – a life cannot be built entirely on faith or on facts. As Buddhists attest, "we will never reach the Pole Star, but we can all walk towards it at our own pace". Everyone sets their own personal pointer somewhere along the scale, and it will move as life goes on. The notional existence of absolute extremes does not make them real, just useful concepts.

Men become civilised, not in proportion to their willingness to believe, but in proportion to their readiness to doubt.

H. L. Mencken

Yuval Noah Harari, in his book *Sapiens*, attributes our ability to entertain Fideism, even in the face of evidence to the contrary, to the power of human language. To paraphrase his thinking:

"Legends, myths, gods and religions appeared for the first time with the Cognitive Revolution 70 000 to 30 000 years ago. The ability to speak about fictions, the things that don't really exist, is the most unique feature of Sapiens' (human) language. We are literally able to believe six impossible things before breakfast – and do so collectively. We can weave common myths such as the biblical creation story, the Dreamtime myths of Aboriginal Australians and the national myths of modern states. Such myths give us the unprecedented ability to cooperate flexibly in huge numbers, with countless numbers of people we've never met. That's why we can live in a dual reality. On the one hand, the objective reality of rivers, trees and lions; and on the other hand, the imagined reality of gods, nations and corporations."

The implication of this way of thinking is that to succeed in life one needs a healthy combination of both. What Harari has done is to remind us that even in the 21st century we are not immune to believing in myths – be they rumours, corporate slogans or advertising campaigns.

Almost all states of knowledge contain a mixture of tradition and new 'facts'. The reality is that those who are flexible in the face of new evidence, are likely to be at the *rational* end of this spectrum.

It has also been noted that the giving up of a belief because it has proven unsatisfactory in real life, or indefensible against new facts or rational data, may actually strengthen the remaining beliefs, no matter how indefensible.

As we know, scientific knowledge is cumulative, building constantly and relentlessly on new research whereas religious authority, by definition, is not. And, with perhaps no surprise to anyone, science keeps pouring out findings that may be uncomfortable for the establishment.

You believe in something if you behave as though it's true.

Alan Watts

Myths are stronger than anyone could have imagined.

Yuval Noah Harari

Natural science revolutionaries in the 18th & 19th centuries

The dominant form of social organization worldwide for many thousand years was an autocratic monarchy and a militarized aristocracy. This fitted in nicely with the religious view of the universe – the king as God, the enforcers as his blessed lieutenants. Might was right. Rules were good.

Similarly, God was simply portrayed as the *Ideal Human* – he looked like one, spoke like one and acted like one. Just a very old, wise, strong and sometimes kind man, a sort of infallible Cosmic Grandpa, a benevolent dictator who set the rules of faith and behaviour.

However, during the 18th century, a new intellectual and philosophical movement invaded the world of ideas in Europe – *the Age of Enlightenment* – promoting concepts such as *scientific rationality* (espoused by Bacon, Descartes and Locke amongst others). Add to this the demonstrable superiority of *collaborative social models* such as corporations and banks in industrialising European societies, and you can see how new thinking chipped away at the old hierarchical and religious structures of previous centuries.

The Age of Enlightenment *was driven by scientific rationality and collaborative social models.*

At the same time interest in natural history was growing with the increasing appearance of strange animals from far-off lands and the constant flood of new fossil finds, from ammonites to the first impressive bones from as-yet unknown giant reptiles. The first Geological Society was founded in London in 1807 and Europe was abuzz with debate around the creation of the earth and the antediluvian world, before *The Flood*.

In this context, let me introduce you to some of my favourite scientific revolutionaries of the 18th and 19th centuries, all of who shaped my romance with palaeontology while living in and around this part of the southern English coastline over the past four decades.

William Smith	James Parkinson	Georges Cuvier	Gideon Mantell	Mary Anning	Charles Lyell	Richard Owen	Charles Darwin
1769 - 1839	1755 - 1824	1769 - 1832	1790 -1852	1799 -1847	1797 -1875	1804 - 1892	1809 - 1882
The first geological map	***Organic Remains of a Former World***	***Essay on the Theory of the Earth***	***The Fossils of the South Downs***	***Important fossil finds***	***Principles of Geology***	***Nautilus & Dinosaur***	***Theory of Evolution***
1801 - 1815	1804 - 1811	1813	1822	1810-1847	1830-1833	1832, 1842	1859
'The Father of Geology'	Study of the antediluvian world 'without concealing any conclusions'	New views based on fossil remains. Established extinction as a fact	Initiated the scientific study of 'dinosaurs'	Changed scientific thinking about ancient life	Established Earth and the universe as billions of years old	Described Nautilus and named Dinosauria	All life evolved from more primitive forms

In the first days of the 19th century, Englishman William Smith embarked on a venture that would result in the creation the world's first geological map. His experience in coal mines and his subsequent observations of the ammonites and other fossils exposed by canal excavations throughout the UK, proved crucial to the formation of his theories of stratigraphy. Smith drew maps from his observations showing the rock types, each with its own colour. Then he estimated the boundaries of each of the outcrops of rock and ended up with a rather unusual map.

William Smith is widely regarded as the 'Father of Geology'.

In 1801, Smith drew a first rough sketch of what would become *The Map that Changed the World* – the world's first proper geological map, finally published in 1815, which inspired the popular book of that name by Simon Winchester.

Across the Channel in France, Georges Cuvier's *Essay on the Theory of the Earth*, published in 1813, provided a new account of the formation of the earth, the geological forces at work on it and the conclusions that could be drawn from the examination of fossil remains. Cuvier was based at the Paris Museum of Natural History and was able to compare the fossil bones he dug up with those of animals alive today.

Cuvier established extinction as a fact – previously considered to be mere speculation.

Guided by the principle of correlation, that all the parts of an animal must be *coherent* and analogous with living species, Cuvier boldly reconstructed extinct creatures from the incomplete skeletons he unearthed and was instrumental in establishing the field of comparative anatomy. Cuvier's work is today considered the foundation of vertebrate palaeontology, and he expanded Linnaean taxonomy by grouping classes into *phyla* and incorporating both fossils and living species into the classification.

Cuvier is also known for establishing extinction as a fact – at the time, extinction was considered by many of his contemporaries to be merely speculation.

Smith's famous 1815 geological map of part of Great Britain.

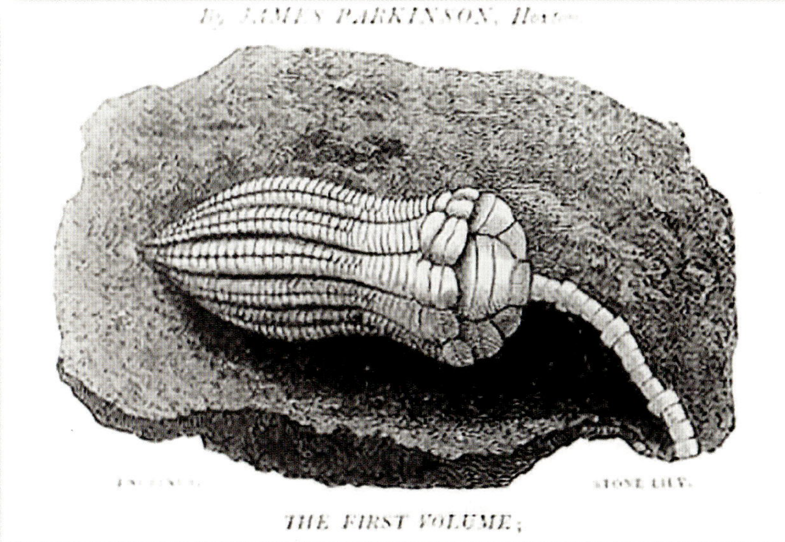

Fossil crinoid illustration from
James Parkinson's *Organic Remains of a Former World.*

At around the same time, James Parkinson, an English surgeon and apothecary best known for his *An Essay on the Shaking Palsy* (now known as Parkinson's Disease), was travelling through mainland Britain with horse and coach indulging in his second love (some say his real love) of palaeontology. He published three volumes on his expeditions between 1804 and 1811, titled *Organic Remains of a Former World.* that reflect the historic conflict between religious dogma and his personal observations from finding fossils of extinct sea creatures high above sea level. He was surprised that so many fossils were easy to identify as they were almost identical to their modern-day counterparts, though some today are only found in tropical regions. He vowed "to conceal no conclusion, however repugnant to popular opinion or prejudice".

After the Deluge: Frontispiece from
James Parkinson's *Organic Remains of a Former World.*

The frontispiece of the book is perhaps meant to be a little ironic: it is titled *After the Deluge* and shows a shattered antediluvian world with Noah's Ark resting on Mount Ararat in the background. Perhaps it heralds a new dawn for thinking about the ancient world.

I will conceal no conclusion, however repugnant to popular opinion or prejudice.

James Parkinson on his research into antediluvian world

Parkinson was, perhaps not surprisingly, one of those who championed William Smith's cause at the Geological Society. He described some fossil species for the first time and illustrated his book with beautiful tinted engraved plates – at a time when the word *geology* was not yet in common use and the word *palaeontology* had not yet been coined.

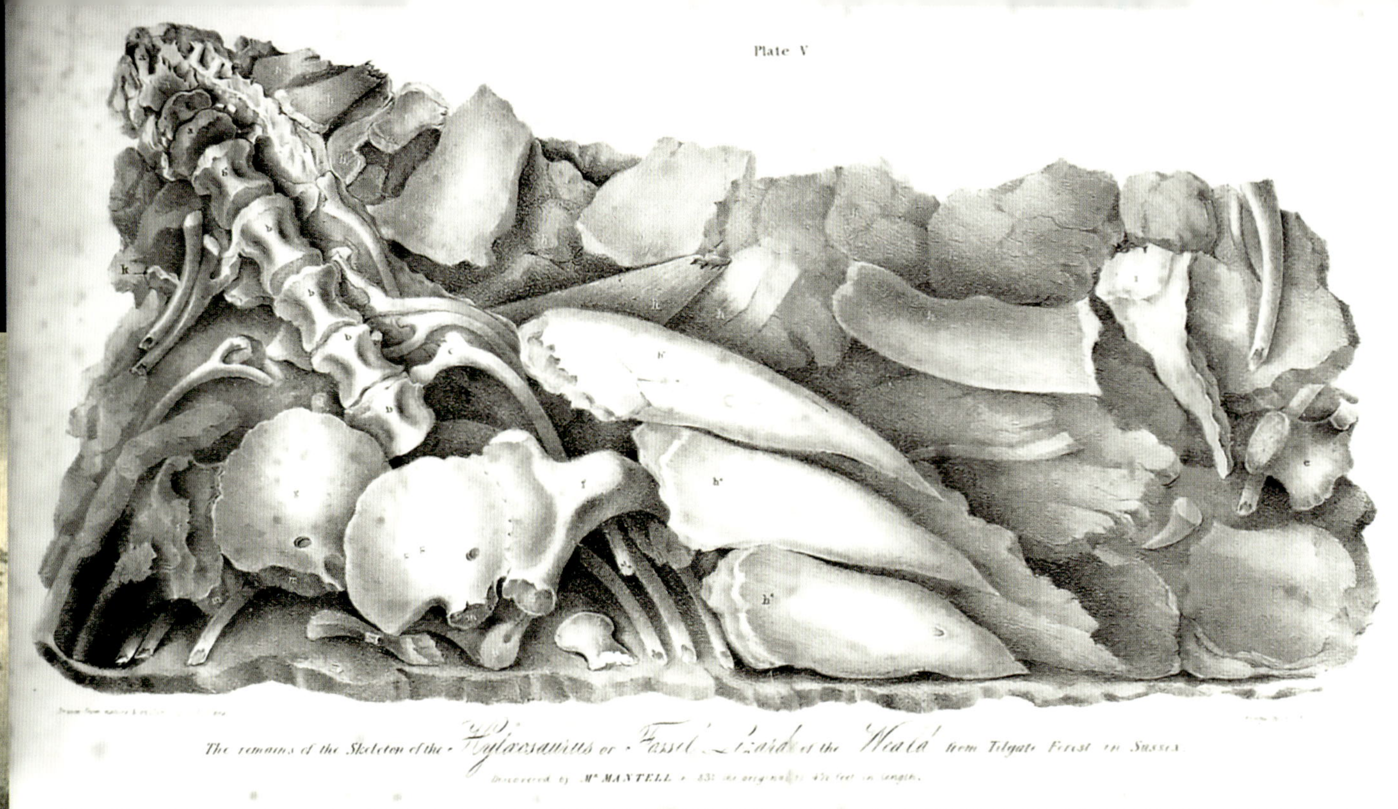

Selected plates from Mantell's 1833 book *The Geology of South-East of England* showing the strata of Tilgate Forest in Sussex England with Gideon Mantell overseeing the collecting of fossils, and some of his bone finds.

Mantell literally began the scientific study of dinosaurs.

Gideon Algernon Mantell was someone who appreciated Parkinson's work. In 1822, he was responsible for the discovery (and the eventual identification) of the first fossil teeth and much of the skeleton of an *Iguanodon*, a genus of ornithopod dinosaur – long before the term *dinosaur* was coined by Richard Owen in 1842. Mantell literally began the scientific study of dinosaurs.

His first (and some consider his most important) geological monograph, *The Fossils of the South Downs; Or Illustrations of the Geology of Sussex*, published in 1822, and of which only 500 copies were printed, contained 42 finely engraved plates of fossils from Cretaceous and Tertiary strata that pay homage to Parkinson's superb images of fossils. The engravings were executed by Mrs. Mantell, from drawings by the author. Mantell also included Parkinson's plates in his *Pictorial Atlas of Fossil Remains* published in 1850.

In November 1810, in Lyme Regis on the south coast of England, an eleven-year-old girl lost her father and continued his passion for fossil collecting in order to supplement the family income, quite unaware of the scientific hubbub in the country. Mary Anning became legendary for the important fossil finds she made in the cliffs around Lyme Regis, the specimens she supplied to academics of the day, many of whom were clergymen, and how these changed scientific thinking about ancient life. As a woman she was excluded from the formal scientific institutions of the time and it was only in 2010, one hundred and sixty-three years after her death, that the Royal Society included Anning in a list of ten British women who have most influenced the history of science.

The 1842 Gray portrait of Mary Anning ready to go fossil hunting with her dog Tray.

Although eclipsed by Mary Anning's fame, another contemporary English woman, Etheldred Benett (1776–1845), devoted much of her life to collecting and studying fossils in nearby Wiltshire. Her fossil collection was considered one of the largest at the time and was for a long time thought to have been lost. However, it was recovered and most of it is now at the Academy of Natural Sciences of Drexel University in Philadelphia. Like Mary Anning, Benett was excluded from the formal scientific institutions but she did work with some of the most prominent geologists of the time,

Etheldred Benett's ammonite chamber specimen as illustrated in the 1836 Bridgewater Treatises, and the actual specimen now in the Academy of Natural Science in Philadelphia.

including Gideon Mantell. Scores of her specimens were included in Sowerby's *Mineral Conchology*, and Buckland illustrated her spectacular specimen showing ammonite chambers in his 1836 contribution to the *Bridgewater Treatises*. In keeping with her anonymity there remains no known portrait of her.

Following in the august footsteps of the luminaries we have met so far, the controversial Richard Owen (1804–1892) was an outstanding naturalist and had a remarkable gift for interpreting fossils. He produced a vast number of scientific works, including the description of the first Nautilus animal in *Memoir on the Pearly Nautilus* in 1832, but is best remembered for coining the word *Dinosauria* in 1842 and giving future generations the word *dinosaur* (meaning *fearfully-great lizard*) to conjure up all sorts of terrifying experiences. He was a vocal critic of Charles Darwin's theory – not in thinking it wasn't valid, but that it was more complex than Darwin thought. Owen's concerns are being mirrored in some of today's new thoughts on evolution.

A young Richard Owen pictured with his Nautilus specimens.

Owen also succeeded in his campaign for the natural specimens in the British Museum to be given a new home with the establishment of London's Natural History Museum in 1881.

What's interesting about the insights garnered by these selected luminaries is that many of them could not bring themselves to admit the full implications of their own,

The more original a discovery, the more obvious it seems afterwards.

Arthur Koestler (1905–1983)

and their colleagues', discoveries. The new *scientific facts* that they were revealing conflicted with the world that they already thought they "understood".

Charles Lyell, who was the first really to glimpse this new review of Creation and evolution, fully embodied this dilemma. His *Principles of Geology*, published in three volumes in 1830–1833, formed the foundation of the scientific belief in a universe billions of years old. Despite this, Lyell had trouble accepting some of the profound implications of his own work, and some then current evolutionary theories. In fact, Darwin's concept of gradual evolution lacked an acceptable time frame until he embraced Lyell's old-earth theories.

In 1859, just a quarter century after Lyell's first publication, Darwin published *The Origin of Species*, famous for introducing his *Theory of Evolution* which proposed that all life, including humans, had evolved from more primitive forms.

Many people saw it as a further attack on religion, because the book made it difficult to believe that God created the world in seven days. It felt as if they had to choose between the two. Some even believed that science had become "dangerous" and was "meddling in matters which only God had control over".

During the second half of the 19th century and the early 20th century, similar expansion in paleontological thinking and the concomitant conflicts with traditional theological teachings spread around the world.

It became apparent that the creation of the earth must have taken place far before that of man. The theological perspective nevertheless continued to develop along the theme that both were acts of creation, and that all living things before man must also have been as a result of an act of creation.

Despite increasing proof from fossils and the science of natural history, the pillars of theology remained largely unshaken. The uniqueness of the human being (evidenced by our apparent *morality*, the so-called *Divine Quality*) and the concepts of the *Flood* and the *Creation* remained centre-stage to many until well into the 20th century and continue in much of humanity today.

Those who believe in *Creation Earth* simply don't believe that life could have existed millions of years ago, despite increasingly clear fossil evidence to the contrary.

It is not the strongest of the species that survives, nor the most intelligent. It is the one that is the most adaptable to change.

Paraphrased from Charles Darwin

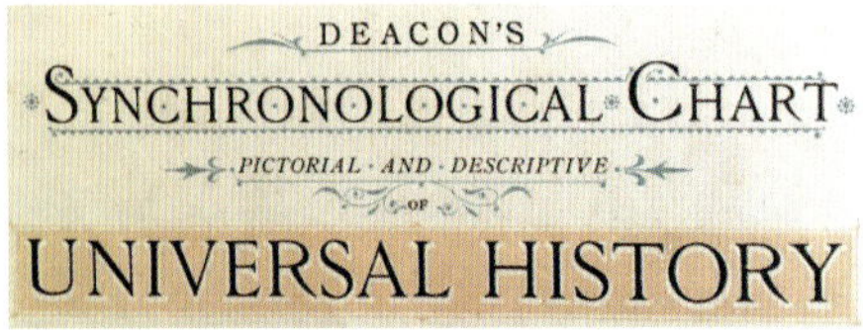

One late-19th century work that tried to document the biblical story of human life on earth, and did so spectacularly, is the *Chronological Chart of Ancient, Modern and Biblical History* originally published by the Presbyterian missionary Sebastian C. Adams in 1871. It is an impressive folding chart, more than seven meters long, that graphically depicts the history of mankind from 4004 BC, the biblical beginning of man, together with a secular history stretching through the Greek and Roman empires to modern times.

Recently I acquired a copy in a local auction, originating from the library of the Pitt-Rivers family, and it has been an item of constant delight and wonder. It is a slightly later edition with the updated and even more ambitious title *Synchronological Chart of Universal History*.

This edition of the chart uses Bishop Usher's chronology, beginning with Adam and Eve and ending on its final page with a stratigraphic timeline that makes no reference at all to the age of the geological sections and how they contradict the start point of this map at just 6000 years ago.

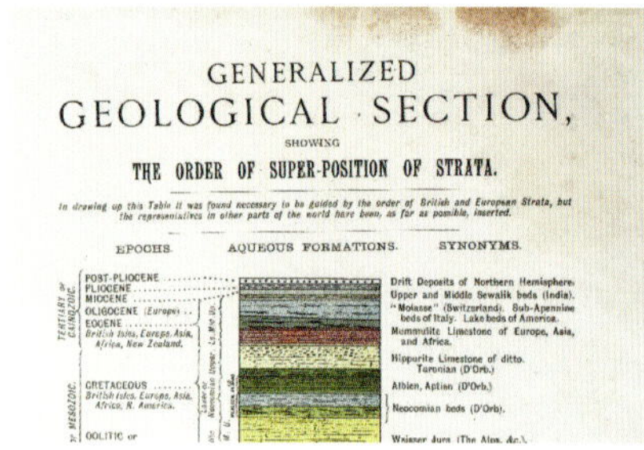

The *Chronological Chart of Universal History* remains for sale in many new editions and lives on both as a serious educational tool, and as a 19th century artistic curiosity.

There are many other creationist texts available today that have now come to accept the fossil record but interpret it as proof that evolution did not happen. They say that because living versions of fossil animals remain today it shows that they were created "perfectly by God" and never evolved.

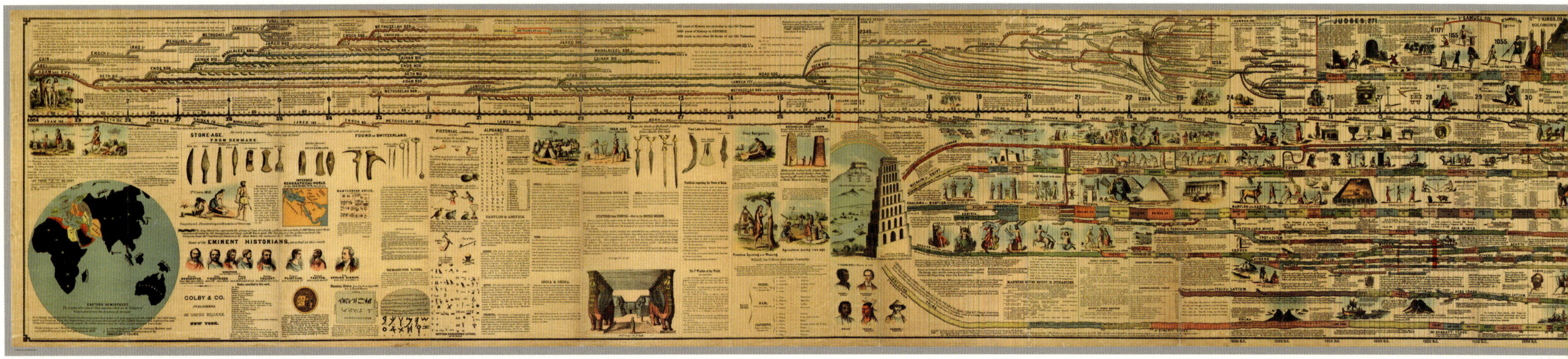

In discussing the major continuities of life within the context of this book, there is always a danger that creationists will leap upon these examples as evidence that *species did not evolve at all*. I was recently handed one such publication, *The Atlas of Creation*, a four-volume tome of thousands of lavishly illustrated pages that compare fossil animals with similar living relatives. Based on this the author comes to the conclusion that all species were created perfectly at a single moment in time. It states that evolution did not take place at all and that "Darwin's evolution is a deceptive, anti-spiritual philosophy and a dark liaison with such bloody ideologies as fascism and communism".

This is a classic example of how selective use of facts can lead to completely fallacious conclusions: molluscs of course did evolve from simple animals found in the Ordovician fossil record but today there are more than 90,000 species described in the fossil record, as well as more than 85,000 highly diverse living species– with new species are being described every month. Just one brilliant example of constant evolution that this text seeks to deny.

This and similar writings skirt around the obvious extinctions of trilobites, dinosaurs and ammonites exactly because their doctrine does not allow for extinction.

The quest for rationality continues unabated.

"These fossils lived 200 million years ago and have never changed. They did not evolve but were created. They have had the same characteristics since the moment they were created. No living thing has ever experienced any transformation. Evolution is invalid."

Harun Yahya, *The Atlas of Creation* 2006

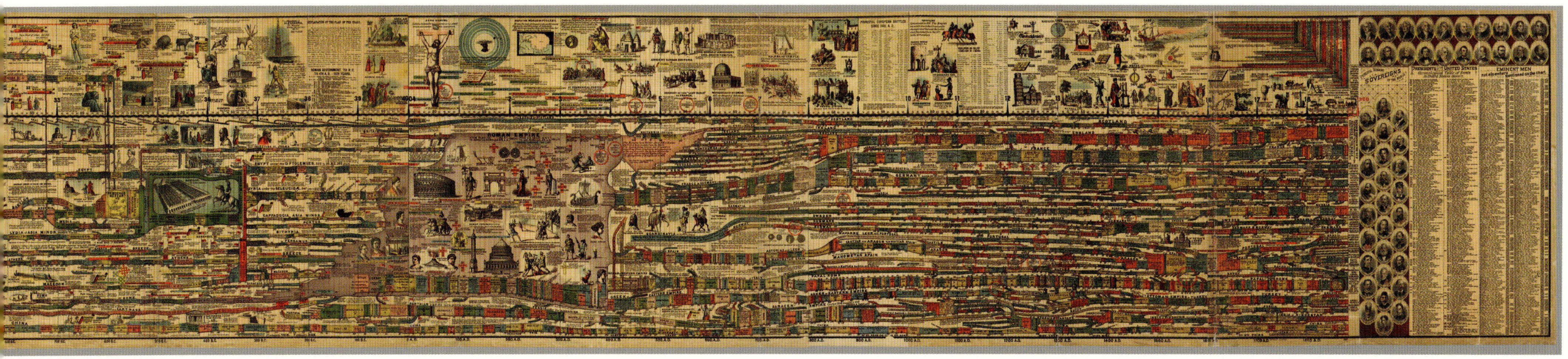

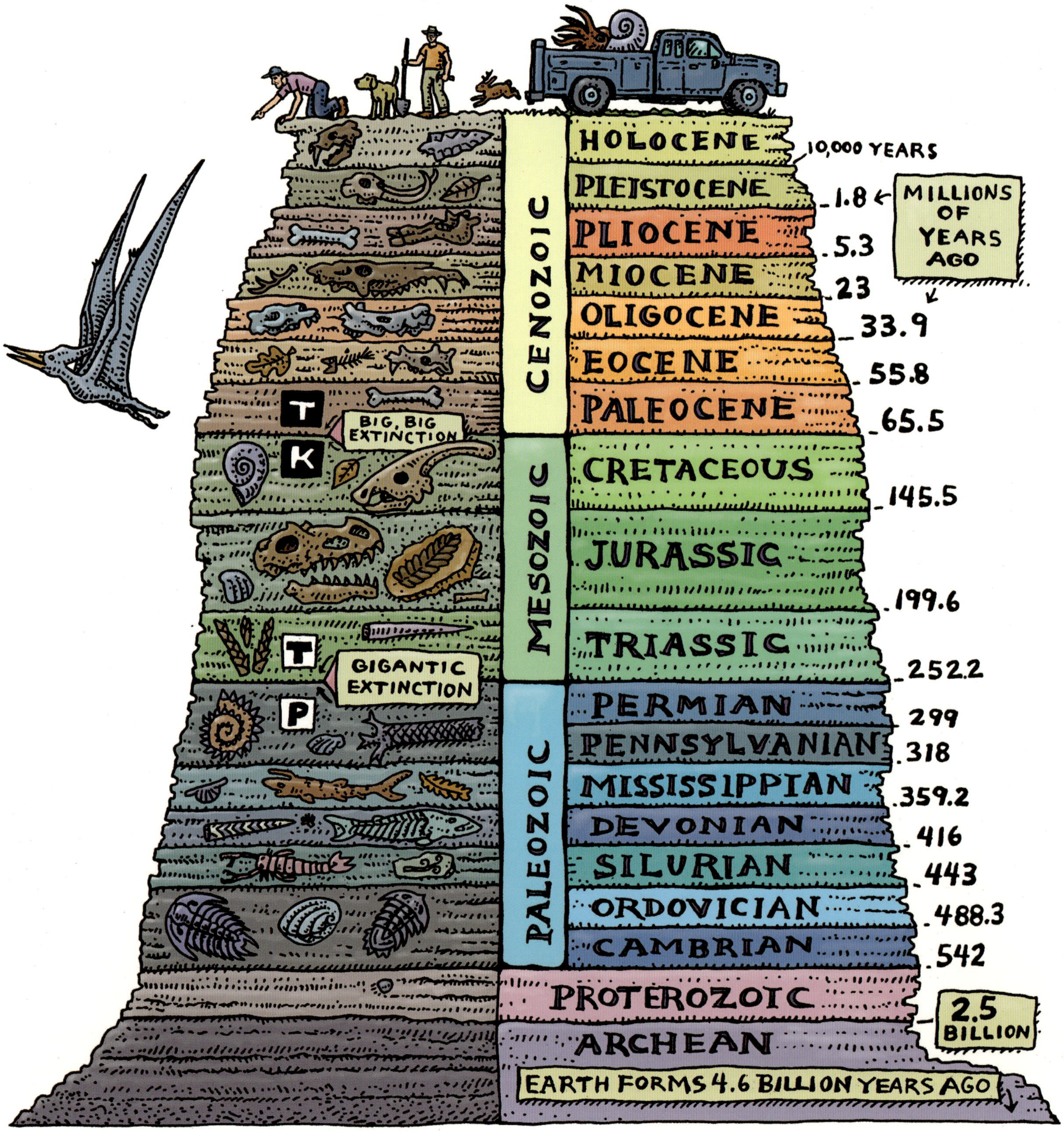

HOLOCENE
10,000 YEARS
PLEISTOCENE
1.8
MILLIONS OF YEARS AGO
PLIOCENE
5.3
MIOCENE
23
OLIGOCENE
33.9
EOCENE
55.8
PALEOCENE
65.5
CENOZOIC
T
K
BIG, BIG EXTINCTION
CRETACEOUS
145.5
JURASSIC
199.6
TRIASSIC
252.2
MESOZOIC
T
P
GIGANTIC EXTINCTION
PERMIAN
299
PENNSYLVANIAN
318
MISSISSIPPIAN
359.2
DEVONIAN
416
SILURIAN
443
ORDOVICIAN
488.3
CAMBRIAN
542
PALEOZOIC
PROTEROZOIC
2.5 BILLION
ARCHEAN
EARTH FORMS 4.6 BILLION YEARS AGO

Geology: the science which deals with the physical structure and substance of the earth, their history, and the processes which act on them.

Dictionary definition

Geology & palaeontology: Enlightened rationalism

The ancient Greeks already studied the composition of the earth, but the word *geology* was apparently first used by Ulisse Aldrovandi in 1603 and introduced as a fixed term by Horace-Bénédict de Saussure in 1779. As we learnt earlier, William Smith drew some of the first geological maps and began the process of ordering rock strata by examining the fossils contained in them, and for this superbly practical contribution is often referred to as *the Father of Geology*.

The science of palaeontology became established in the 18th century as a result of Georges Cuvier's work on comparative anatomy, and developed rapidly in the 19th century, while the term *palaeontology* was first published by the editor of a French scientific journal in 1822. Unlike the study of plants and animals in the present-day world, palaeontology introduced the dimension of time to the classification of life.

It had become clear to everyone that many rocks had layers that were piled atop one another and, given the known processes of sedimentation, it was logical to infer that each layer was older than the one above it. Some clearly identifiable fossils – ammonites, for example – were found in lower layers but then disappeared at a certain point below the surface. This clearly implied that they had vanished from the face of the earth some time before the present.

That animals could become extinct naturally and without human agency was a controversial proposal in the early nineteenth century. Sensibly seeking to avoid directly contesting the social orthodoxy of the time, fossil specialists allowed their subject to be viewed as the study of the "ancient" or "prehistoric" world as if that were a separate planet from the one they now inhabited.

"The world was very different back then" remains a common theme in old books and museum exhibits of fossils. However, this separation of *history* from *prehistory* has no scientific validity whatsoever. As we will see in further pages, we live today among a broad mixture of young and old bloodlines, many of which stretch more or less unchanged back to the beginnings of life as we know it.

The system of *life* is thus best portrayed not as one broad river of time interrupted by scattered dams but as thousands of separate streams of wildly differing volumes, speeds, widths and lengths. As we have already seen, our notion of *time* is merely a

Palaeontology: the branch of science concerned with fossil animals and plants.

Dictionary definition

Deep history is always being rewritten: if it were not, there would be no science; there would just be a long list of facts.

Richard Fortey in the Foreword to his book Life in 2008.

Biology: the branch of science concerned with the study of living organisms.

Dictionary definition

consensus as to how we measure its passage. Each of the myriad streams has its own rate of evolutionary change and the effective pace of time thus greatly differs for each one.

How useful it would be to study all animals – the ancient and the contemporary – as a single stream. A crab in Jurassic limestone and a crab in a jar of ethanol are both crabs, first and foremost.

If, then, one aims to study (for example) crabs, how much more useful it would be to study all crabs – the ancient and the contemporary – as a single stream. A crab in Jurassic limestone and a crab in a jar of ethanol are both crabs, first and foremost; in terms of their evolution, the fact that one is preserved in stone and the other in fluid is a meaningless distinction. By investigating equally all crabs that have ever been, one comes to understand what crabs really are and perhaps even what crabs can be.

We need to find a term for the complete study of life, across all of time. We could for example drop the *paleo* from *palaeontology* and simply call it *ontology*. However, the term has already been claimed by a branch of philosophy called metaphysics and incorporates *the study of being, becoming, existence and reality*. So, we need to find another term for our purpose.

Why not just settle on *Biology*? Defined as "the study of living organisms", this definition appears to ignore the entire history of living organisms, as well as the potential futures of these. It's almost as if there is nothing to learn from the past, or from the future.

As we take a top-down view of life in the ocean, it is clear that we really need a term that is more inclusive, one that encompasses the whole continuum of life, to encompass biology, palaeobiology and palaeontology.

What could that term be? The jury is out.

Phylogeny: A big word for family tree

Many palaeontologists will tell you that they have absolutely no problem putting across hundreds of millions of years of life's history, and to make it comprehensible to their audiences. Deep time, they say is easy to comprehend, and communicate.

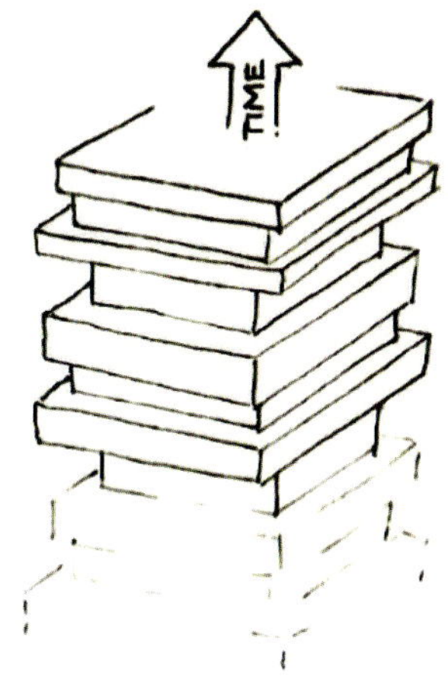

The most common way to illustrate geological time is with a series of sequential horizontal epochs, each one reflecting a period during which strata were deposited together with their fossil remains, each one younger than the one below it, and then set against a vertical timeline.

This stratigraphic approach has been universally popular – the illustrations below are examples from Cuvier (1834) and Gregory (1951).

Already in Gregory's 1951 image you can see an attempt to integrate the *strata view* with a simplified tree structure to highlight relationships between animals. The thinking seemed to be that if you slap a branching *tree-of-life* on top of the *strata-view*, then everyone will understand the evolution of various life forms over time. QED.

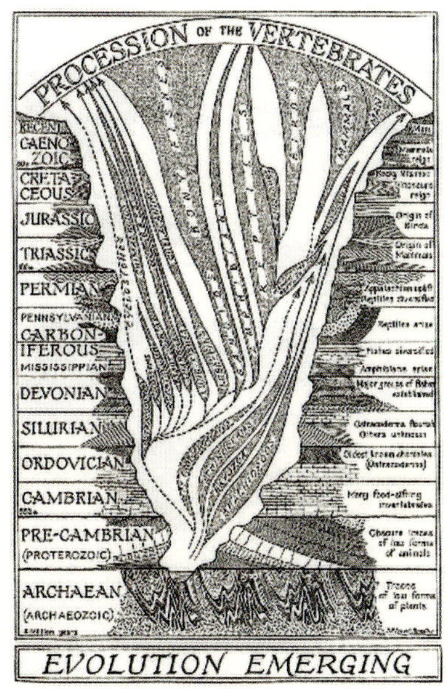

An Epoch-based Geologic Time Scale and Mass Extinctions

When viewing these epoch-based views of evolution, remember that sudden changes happen when there is no extinction event and even when there are no dramatic changes in the environment. We need to be able to view the entire history of life over time – without epochal boundaries – and celebrate its surprising continuity.

Era	Period	Major events
Cenozoic	Quaternary	Evolution of modern humans
	1.8 mya	
	Tertiary	Mammals and birds diversify into space left by dinosaurs
	66 mya	
Mesozoic	Cretaceous	Extinction of non-avian dinosaurs Extinction of ammonites First primates First flowering plants Heteromorph ammonites diversify
	146 mya	
	Jurassic	Dinosaurs diversify First birds evolve from theropod dinosaurs
	200 mya	
	Triassic	First mammals First dinosaurs First true ammonites
	250 mya	
Paleozoic	Permian	Reptiles diversify First modern trees: conifers, ginkgoes & cycads.
	299 mya	
	Carboniferous	First four-legged vertebrates First reptiles Huge forests & swamps create coal deposits
	359 mya	
	Devonian	First insects & spiders Fishes diversify: bony fishes & sharks First amphibians First seed-bearing plants
	416 mya	
	Silurian	First vascular land plants
	444 mya	
	Ordovician	Diversification of trilobites and cephalopods Barren continents
	488 mya	
	Cambrian	First trilobites First primitive molluscs & cephalopods First jellyfish, echinoderms & primitive fish
	542 mya	
Late Proterozoic	Pre-Cambrian	First soft-bodied metazoans First skeletal elements Stromatolites First traces of animals
	650 mya	

This evolutionary or lineage-based view using a waxing, waning and branching tree structure led to some of the most intricate images ever seen, when they first appeared in the mid-sixteenth century. The proper name for these classic *Tree-of-Life* diagrams is a *Phylogeny*.

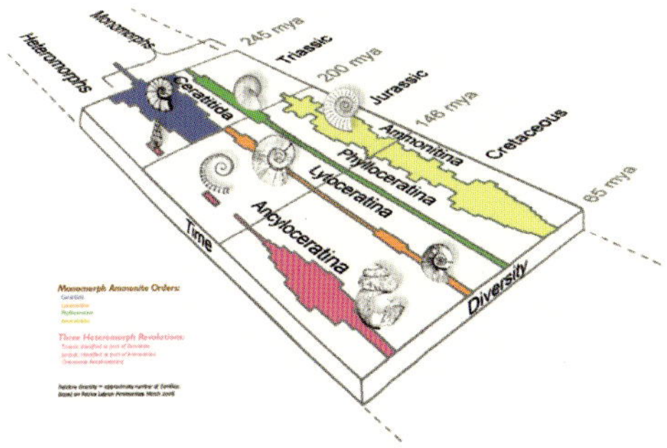

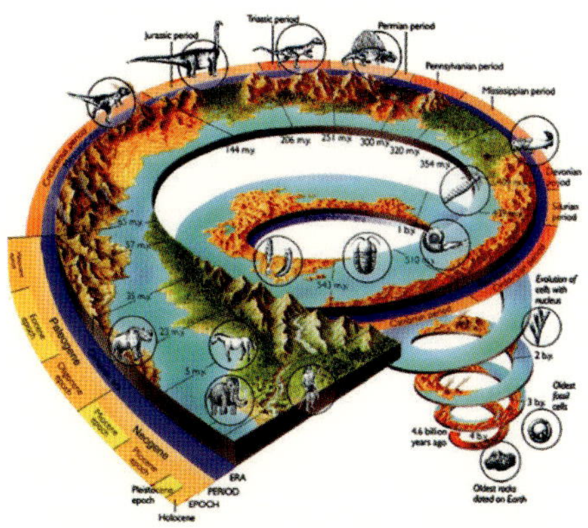

All of these representations are useful in their own way, and I have used them, and others, in the two previous books in this trilogy. In various discussions over the past few years I realised that each introduces subtle implications. In all strata-based views (for example such as the large image on the facing page) there is an implicit suggestion that evolution is strongly affected by epochal changes – which does happen, true, but just as often it doesn't. Strong and sudden evolutionary changes happen even when there is no *extinction event* or even when there is no dramatic change in the environment.

What has become most disturbing to me is that none of these highlight the powerful continuities that exist in the history of life.

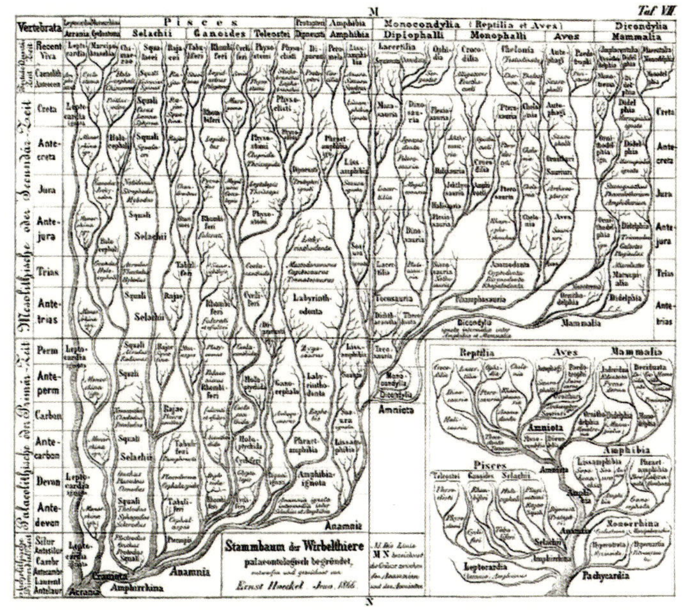

This 1866 image by Ernst Haeckel – his version of the vertebrate family tree – is one of the great examples of a classic Tree-of-Life.

A conceptual phylogeny of the continuity of animal life

Just because it's difficult to communicate the story of *the-tree-of-life* doesn't mean to say that we shouldn't do it or that we won't. So, in the next chapter, which starts overleaf, I give you my interpretation of the-tree-of-life, my own *conceptual phylogeny* of animal life. It's focused on telling the story of the amazing continuity of life in, out of, and around the ocean.

I started out with a focus on how life began in the ocean, questioning why and when life moved into the water column, eventually emerged onto land, ventured into the skies, however intermittently, and then evolved back into the ocean again, and again and again.

I hope that my *conceptual phylogeny* will clarify and inspire, rather than add another level of noise to the temporal fog. At the very least it will help to connect the strands and stories in this book.

Phylogeny: the history of evolution among broad groups of organisms, showing lines of descent and relationships.

Dictionary definition

THE CONTINUITY OF ANIMAL LIFE

A Conceptual Phylogeny & Family Tree

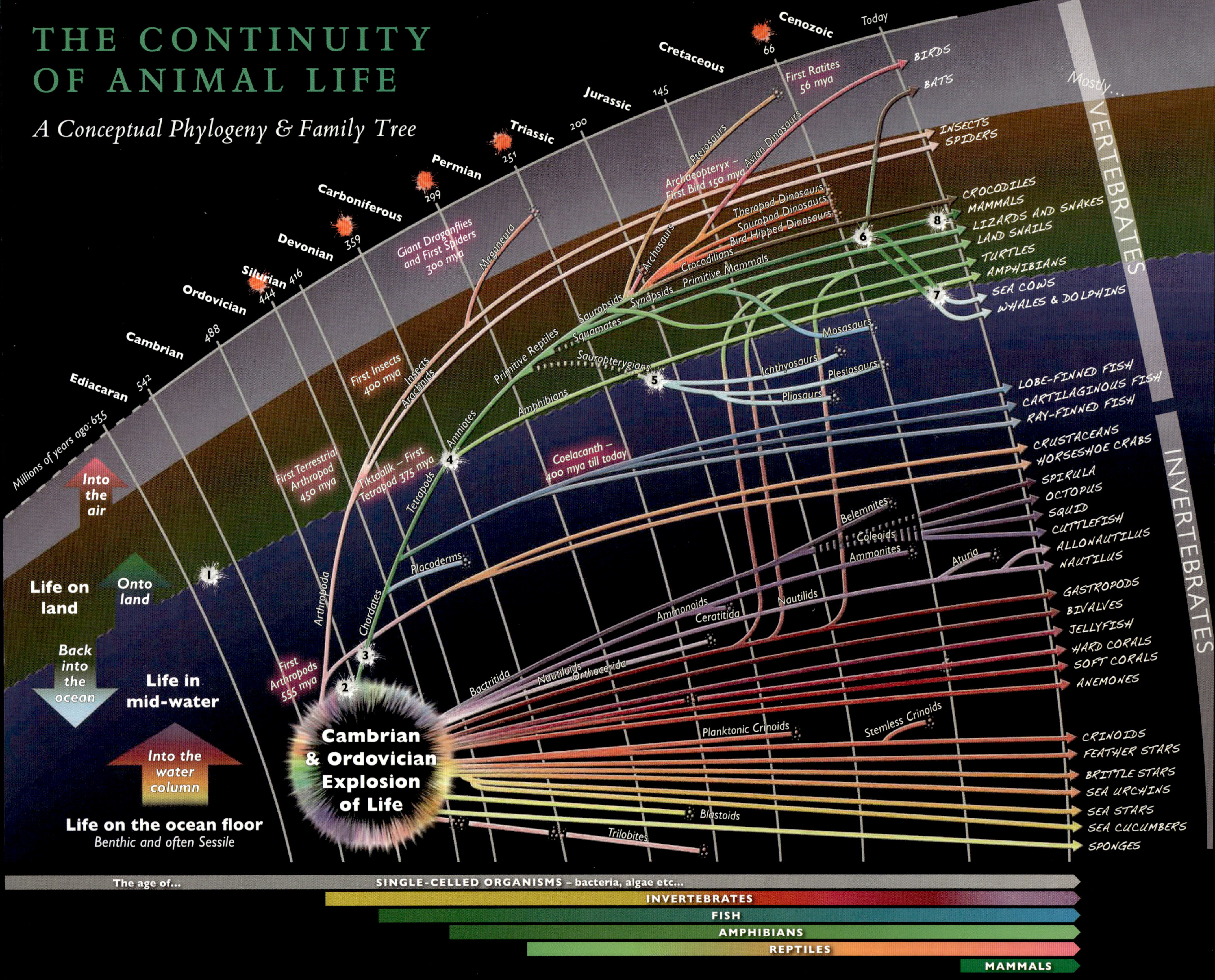

- Mass extinctions
- Group extinctions
- 'White Space' for new life

1. Plants make land habitable
2. Evolution of shells and hard parts
3. Chordates: Evolution of skull and vertebra
4. Amniotic egg
5. Competition on land. Abundance of fish
6. Absence of predators enabled mammals to thrive
7. Abundance of fish encourage mammals to enter the sea
8. *Homo sapiens* emerges

EVOLUTION & EXTINCTION

A 500-million-year experiment

Fossils are the way we know about deep time. Without fossils we would know little about ancient life and we would be much more likely to believe in fables and myths to fill the gaps in the vast expanse of time. There would be little else that we could rely on to fill that void without fossils. We would know little about the 500-million-year story of complex life.

The study of fossilised life, palaeontology, is conventionally based on a linear view of time that divides the past into named epochs – the Permian, Jurassic etc. This approach favours the view of life within time slices, rather than that of related groups (such as reptiles or fish) throughout time. The reality though is that many of these epochal phases may have gone completely unnoticed by many forms of life on earth – such as corals, which lived across many epochs and extinctions.

In my view, the pathways of life, evolution, extinction and continuities are thus better understood if we step back to get a broad perspective. For this reason, I created this chart that highlights the continuity of life – it's a simple, visual *Conceptual Phylogeny of Life* – a high-level *Tree of Life* that chronicles the major forms over time but adds one other important element – their mode of life. Did these organisms live a sessile life on the sea floor, live in the water column, move onto land or venture into the air? When and why did some land animals return to live in the sea?

Notice how many of the timelines of the major animal groups represent major continuities and how relatively few of them end in extinctions.

This chart represents five hundred million years of evolutionary relationships on one page. The first thing that should strike you is that how many of the horizontal timelines stretch across the epochal boundaries, right until today, and how few of them end abruptly in extinction.

Despite their usefulness, fossils are limited in what they can tell us about the past because typically only the hard parts of animals are mineralised. Darwin was well aware of fossils and their natural limitations as a source of evidence and in 1871 he wrote to Joseph Dalton Hooker, a botanist and his closest friend, about his desire to find "some warm little pond" that would contain what came to be known as the *primordial soup* of life – some evidence beyond fossils.

Beware of vernacular common names that over-generalise. A 'snail' may sit on a rock next to another 'snail' from a completely different branch of the gastropods that emerged from the ocean millions of years before. That's like calling all humans and chimpanzees 'humans'.

Modern-day biological groups (for example *fish*) are artificially constructed groupings of similar species that are derived from a common ancestor – but not necessarily a recent one. Many marine families evolved a very long time ago but have not since changed enough to form new families. Even though snakes have evolved to such a point to have lost their legs completely, they are still included within the tetrapods – four-legged animals.

The phrase "Survival of the Fittest" was first used by Herbert Spencer after reading Charles Darwin's On the Origin of Species *in 1864. It describes the mechanism of* natural selection *and is best understood as "Survival of the form that will leave the most copies of itself in successive generations."*

To gain a deeper understanding of the pace of evolution and the durability of certain body plans we should link the earliest instances of a family to its current ones and seek to fill in the gaps, regardless of the length of time involved – which is what my *Conceptual Phylogeny* tends towards.

Conditions changed dramatically throughout geological time, causing animal species to evolve or to fail in the attempt, and eventually become extinct. This is called the Survival of the Fittest. Sometimes conditions changed catastrophically, on a global scale, causing what we commonly call *mass extinctions*.

Five mass extinctions

Our earth has witnessed five extinction events during which more than 75% of living species disappeared each time. We know this primarily from the fossil record and have now concluded that most loss of life had to do with rapid climate change following spectacular geological events.

The extinction event at the end of the Cretaceous Period remains the most iconic because it wiped out all of the non-avian dinosaurs. Some of the other mass extinctions were even more destructive but have not proved as popular in the media. Here is a summary of the top five:

1. The mass extinction at the end of the Ordovician 444 mya had two peak dying times separated by hundreds of thousands of years.

Cause: Massive glaciations locked up much of the world's water as ice and caused sea levels to drop precipitously, possibly triggered by the uplift of the Appalachian Mountains in eastern North America. Tumultuous weathering and the formation of limestones made of calcium carbonate sucked CO_2 out of the atmosphere, further chilling the planet.

Effect: During this time most life was in the ocean, so it was marine organisms that were drastically reduced in number – such as shelled brachiopods, the eel-like conodonts, and the trilobites. It is estimated that 86% of species were lost.

Teeth that lined the mouth and throat of a conodont, a jawless eel-like vertebrate. Palaeontologists were baffled about the origin of these toothy fragments, mistaking them for bits of clams or sponges. The discovery of an intact fossil conodont in Scotland in the 1980s revealed their owner. These teeth were one of the first structures built from hydroxyapatite, a calcium-rich mineral that remains a key component of our own teeth and bones today.

2. The Late Devonian mass extinction 360–375 mya was likely a drawn-out series of extinctions each lasting 100,000 to 300,000 years, over a span of perhaps 20 million years, rather than a single event.

Cause: The likely cause was the exponential development of land plants during the Devonian period. Their deep roots stirred up the earth, releasing nutrients into the ocean on an unprecedented scale, triggering algal blooms which sucked oxygen out of the water, suffocating benthic bottom-dwellers like the trilobites.

Effect: 75% of species were lost. Life in shallow seas and on reefs was worst affected, not returning to its former glory until new types of hard corals evolved over 100 million years later. The armour-plated placoderms went extinct. Trilobites were the most diverse and abundant of the animals at the time and although they had survived the previous mass extinction, they were nearly wiped out in this second one. Insects, plants, and the first proto-amphibians were on land by then, and these new landlubbers were also dealt a severe blow.

3. The End-Permian mass extinction 251 mya has been nicknamed *The Great Dying*, an unthinkable volcanic nightmare in which life was nearly wiped out.

Cause: Many experts believe an asteroid triggered this mass extinction, but no crater has been found. Perhaps caused by such an impact, a cataclysmic eruption in Siberia produced more than a million cubic kilometres of lava, blasting CO_2 into the atmosphere. Bacteria responded by expelling massive amounts of methane. Global temperatures surged while oceans acidified and stagnated. Fossils immediately after this period include no corals or new coal deposits.

Effect: An astounding 96% of marine species were wiped out, as well as 70% of life on land. Trilobites and blastoids went extinct. The rugose and tabulate hard corals disappeared completely.

4. During the final 20 million years of the Triassic period, there were several phases of extinction which together is called the **End-Triassic mass extinction 200 mya.**

Cause: Floods of lava erupting from the central Atlantic likely led to massive climate change, although an asteroid impact has also been blamed. Of all the mass extinctions, this is the most curious. No clear cause has been identified but it was a period of strong ocean acidification.

Effect: Almost 80% of species went extinct including 20% of all marine families (including the ammonite Orders Ceratitida and the Orthocerida which had by then disappeared from the fossil record), most primitive mammal-like creatures, many large amphibians, and all non-dinosaur archosaurs.

5. The **End-Cretaceous mass extinction 66 mya** (previously known as the K/T extinction, or lately as the Cretaceous-Paleogene K-Pg event)

Cause: A meteorite more than ten kilometres wide struck the shallow sea between the continents of North and South America with a force of 10 billion Hiroshima bombs, creating a crater 180 kilometres across, off what is now Mexico's Yucatán Peninsula. It is marked by a thin layer of iridium (a metal which is rare in the Earth's crust, but abundant in asteroids), a deposit which can be found throughout the world in marine and terrestrial rocks.

Effect: The impact melted the ground several kilometres deep and sent up a plume of hot dust and molten rock creating a chain of events that changed the fate of the planet in seconds. It is popularly called *The Day the Dinosaurs Died*, and what a day it must have been.

A 10,000°C firewall raced in all directions destroying everything in its path. Under darkening skies, this spectacular wall of fire reached Patagonia at the tip of South

America in less than two days, destroying the vegetation and large herds of dinosaurs feeding on it. A tsunami spread out in all directions and reached the coast of north west America, 2000 miles away, in just over two hours, throwing all manner of marine life onto the land.

Within a week the skies were filled with toxic dust, it was dark and almost all large animals lay dead or dying. All of this was exacerbated by ongoing volcanic eruptions in India's Deccan Flats.

If the meteorite had hit anywhere on the planet except here, perhaps somewhere in a deeper part of the ocean, then the consequences would have been different and far less serious. The site of the impact crater in shallow water meant that the chemistry of the atmosphere was instantly changed by the large quantities of gypsum thrown up in the impact plume and distributed world-wide. The sulphates contained in it acted to block out sunlight and resulted in a 10% temperature drop. This *dark and cold winter* lasted for perhaps a million years. The impact turned our blue planet grey. There was almost nothing to eat. Those dinosaurs that did not die in the fires, heat or tsunami impact, soon starved to death. And, so ended the age of the dinosaurs, the dominant vertebrates of the time, and that of the ammonites. This extinction, when 50–75% of all species were lost, has left the most fossil evidence of all the mass extinctions.

This rather arbitrary group of five mass extinctions were caused by asteroids, volcanoes and other external events, but most of the extreme loss of animal life was caused by consequential dramatic climate change that decimated food webs.

In the end it was all about climate change.

From extinction to continuity

More than 90% of all species that have ever lived on earth are extinct.

National Geographic

Reflecting on these mass extinctions you can see that no matter how severe they were, for life in the ocean they were little more than bumps in the road to the future. You can think of them as *evolution interrupted*. Alternatively, you can think of each extinction as *creating a world of new possibilities*.

How often have you heard statements similar to this one "More than 90 percent of all species that have ever lived on Earth are extinct." With five extinctions each destroying more than 75% of species that should not be a big surprise.

90% of all major animal groups that existed in the Jurassic are alive today.

What is surprising is the continuity of life, illustrated by the fact that 90% of the major animal groups that existed in the Jurassic are alive today.

Certainly, some animal groups have disappeared completely (for example trilobites, ammonites, pterosaurs, marine reptiles and non-avian dinosaurs), and forever. But, and importantly, we need to look beyond extinction, to understand the second-order implications of their departure, by learning what new opportunities their absence created and what subsequent life forms capitalised on this new *white space* and replaced them in their niches.

Dinosaurs first appeared after one of the biggest mass extinction events ever, the Permian-Triassic extinction about 250 million years ago. The later mass extinction, at the end of the Cretaceous Period 66 million years ago, killed off the non-avian dinosaurs and made room for mammals to diversify and evolve rapidly.

Mammals thrived in the absence of dinosaurs. We have their extinction to thank for our presence here.

At the time of the extinction of the dinosaurs and the destruction of most of the plant life on land, the largest mammal was the size of a shrew, and these small scavenging animals were able to survive on seeds and nuts. As plants recovered from the dark cold winter after the meteorite impact, mammals evolved quickly into the ecological white space left by the dinosaurs. For them it was a world full of new possibilities. Mammals thrived in the absence of their previous predators.

Think of this process as analogous to a traffic light – each extinction creates white space into which new life can evolve – a constant cycle of redevelopment.

Extinction = Departures

White Space = Innovations

New Life = Arrivals

We need to look beyond extinction, to understand what opportunities for new life that extinction creates.

Evolution, it turns out, is extremely opportunistic – with each new empty niche pregnant with opportunities.

Evolution only happens at a species level – its engine is genetic mixing at the point of reproduction. Evolution is now known to happen at widely differing speeds depending on the organism. This is strongly linked to lifestyle and environment. Animals such as the coelacanth have remained largely unchanged over hundreds of millions of years, while ammonites changed at such geologically breakneck speeds that they are used to date the rocks in which their fossils are found. It's nature's natural chaos.

Body plans that recur repeatedly in various animal groups (for example the similar sleek fish-forms of ichthyosaurs, mosasaurs and dolphins) can be thought of as *optimum adaptations*. The longest-lived groups (for example crinoids, sea urchins, molluscs, nautiluses etc) can be thought of as having achieved *optimum designs* that stood the ultimate test of time.

Among the first iterations of life that we can recognize in the fossil record are the simple single-celled stromatolites we met earlier in the book. They are a variety of beings that formed distinctive shapes and mats on rock surfaces in the water. They span the interval between non-cellular and cellular life and go back more than three billion years. Their vast fossilised layers are commonly found world-wide and they still thrive in a few localities today.

These are aspects of evolution that I want to highlight throughout this book plus the somewhat surprising ways in which life goes on to thrive again. But we are really getting ahead of ourselves here – let's start with life at the beginning, even before the fabled *Cambrian Explosion of Life*.

Life before the Cambrian

Until the start of the 20th century it was commonly accepted that *The Cambrian Explosion of Life* signalled the start of complex life on this planet and that life before this time was nothing more than single-celled organisms.

Fossils from the Ediacara Hills that may be early ancestors of corals, echinoderms or arthropods.

In 1946, Reginald C. Sprigg, an Australian miner, was exploring a range of mountains near Adelaide, Australia, known as the Ediacara Hills. He found fossilized imprints of "soft-bodied organisms" – round, disc-shaped forms that he dubbed "medusoids" due to their similarity to jellyfish. Some were giants, thought to be 1.4 m in length. Others resembled worms, arthropods, or nothing at all. Some think of these as "experiments in animal-like beings". This period became known as the *Ediacaran Biota* and put back the time of recognisable animals, to the Pre-Cambrian period, 635–542 mya. The animals in these fossil deposits lived on the sea floor amongst microbial mats and demonstrated some really weird body plans, some with still highly contested identities.

Fast forward another twenty years to the eastern-most point in North America, called Mistaken Point, named for the shipwrecks caused by this foggy cryptic coastline. Here, in 1967, fossils found here showed imprints of large complex soft-bodied marine creatures – long before multi-celled animals should have existed.

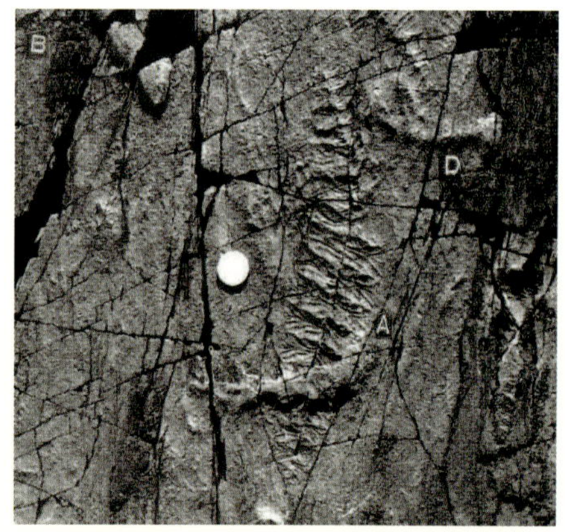

Spindle shaped organism Fractofusus with primary and secondary branches. Mistaken Point, Canada.

The Ediacaran animals were unlike any beings alive today – but they disappeared almost inexplicably before the Cambrian.

Specimens have now been collected from more than sixty Ediacaran sites. These primitive animals did not rely on photosynthesis for energy but, lacking mouths, guts and anus, they drew their nutrients directly through their outer covering from the water and microbial mats. They were unlike any beings alive today – but they disappeared almost inexplicably by the end of the Ediacaran.

Some scientists now believe that "Swiss Cheese" sedimentary layers of the early Cambrian show that an explosion of wormy creatures laid waste to the weird soft-bodied Ediacaran populations. Some scientists have started calling this period *Worm World*. The Ediacaran animals were certainly not *failed experiments* – they exhibited their own kind of continuity for more than 30 million years. Amongst these newcomers must have been the ancestors of sponges, jellyfish, molluscs, sea anemones and corals – biologically complex beings that arrived much earlier than the Cambrian Period.

The Cambrian: An explosive cauldron of life

Anatomical variety reached a maximum in the Cambrian. Compared with the Burgess seas, today's oceans contain many more species, but based on fewer anatomical plans.

Stephen Jay Gould

The Burgess Shale, a series of Middle Cambrian (508 mya) fossil deposits in British Columbia, Canada, is famous for the exceptional preservation of soft-body imprints. It is regarded as the best-preserved record of life more than 500 million years ago and documents the Cambrian revolution of life.

In the Cambrian, there was at last enough atmospheric oxygen to fuel the metabolism of complex organisms. Any that developed had vast supplies of bacteria to feed on and the empty ocean was devoid of competitors.

New life forms emerged with gay abandon. Any forms that could be tried were tried. You only have to look at the creatures preserved in the Burgess Shale fossil beds. There is evidence of life forms so weird that they are impossible to categorise into any of the animal groups we know today.

An animal such as *Opabinia* is likely an arthropod but its features are nothing like any other arthropod ever known. These forms just don't seem to make sense – *Opabinia* still does not fit comfortably into any traditional schema.

The Cambrian explosion of life was a rare bubbling cauldron of innovation, but we now know that the *Explosion of Life* started much earlier than was once thought, and in fact it continued much later than the Cambrian too, in several separate pulses.

For cephalopods this phase of innovation continued well into the Ordovician – a process covered in some detail in the second book in this trilogy, *Nautilus*. The explosion of life was much more a continual process covering perhaps 100 million years than a one-time sudden event.

Opabinia fossil at the Smithsonian Institution.

Opabinia reconstruction.

The explosion of life was more of a continual process, covering perhaps 100 million years, than a one-time sudden event.

THE GREAT INNOVATIONS
Shells, bones, eyes and eggs

While extinctions created white space into which new life could flourish, there were several major innovations that created substantial competitive advantage for life and enabled great diversification. This section explores just a few of those.

The Great Innovations
Life gets hard

Without hard parts and the process of fossilisation, we would have known little about the history of life.

The Cambrian Period is the first time that ancient soft-bodied animals created *hard skeletons* – a radically new concept, long before vertebrates with *internal* skeletons emerged.

Worm tubes, mollusc shells and sponge spicules are examples of earliest hard parts created by soft-bodied animals. The gastropods (the snails) and the bivalves (the clams) were amongst the first to produce hard external shells.

The evolution of primitive bones, skulls and vertebrae 530 million years ago signalled the start of bones and vertebrate animals – the chordates.

The emergence of these *hard parts* in animals signalled the start of fossils. For the next 500 million years, evidence of invertebrate and vertebrate life was to be preserved in hard mineralised forms. Without hard parts and the process of fossilisation, we would have known little about the history of life. Just consider that most dinosaurs are only known from fossil bones, and you have an indication of the importance of hard parts.

Associated with the formation of hard shells in the early cephalopods was the formation of *chambered shells*, epitomised by the nautiloids. Cephalopods with chambered shells were the first animals, after jellyfish, to be able to move off the sea floor into the water column. This gave them a distinctive competitive advantage and spawned spectacular profusion and giant predators – which we'll meet later.

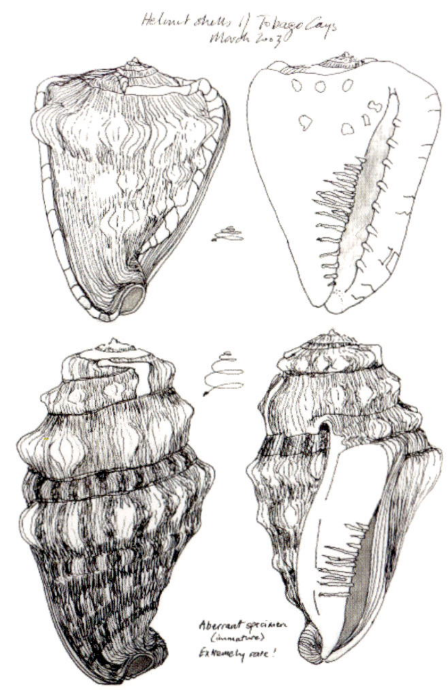

A cross-section of ammonite shells gives us the opportunity to study their internal and shell formation, from the first juvenile whorls to their adult stage, plus the association with other life forms present in their environment at this period in the Jurassic almost 200 million years ago.

The Great Innovations

The eyes have it

Marbled Stargazer *Uranoscopus bicinctus* photographed in Papua Province, Indonesia.

In the Pre-Cambrian, animals were simple and so any light-sensitive organs they had, were primitive. Simple light detection is found in bacteria, single-celled organisms, plants and animals.

The basic biochemistry on which vision relies is common to all eyes, but they have a wide range of structures and forms, all of which have evolved quite late relative to the underlying proteins and molecules. Photoreceptor cells may have evolved more than once, and functional complex eyes seem to have evolved independently several times from these. Each form of eye is fine-tuned to the requirements of its owner; with varying acuity, wavelengths it can detect, ability to detect motion and colour sensitivity.

Prey generally have eyes on the sides of their head so to have a larger field of view, in order to avoid predators. Predators, however, have eyes in front of their head for better depth perception. The stargazer, a fish that prefers a sedentary life under the sand, has eyes on top of its head and a large, upward-facing mouth. It will leap upwards out of the sand to ambush prey that passes overhead. Because of stargazer ambush strategy and its ability to deliver both venom and electric shocks, it has been called "the meanest thing in creation".

Complex, image-forming eyes have evolved independently several times and the variety of solutions that evolved, even in related species such as cephalopods, is astonishing – as the comparison of octopus, cuttlefish, squid and nautilus eyes shown in the images below. Eyes and other sensory organs probably evolved before the brain and provided essential inputs that changed perception, lifestyle and success.

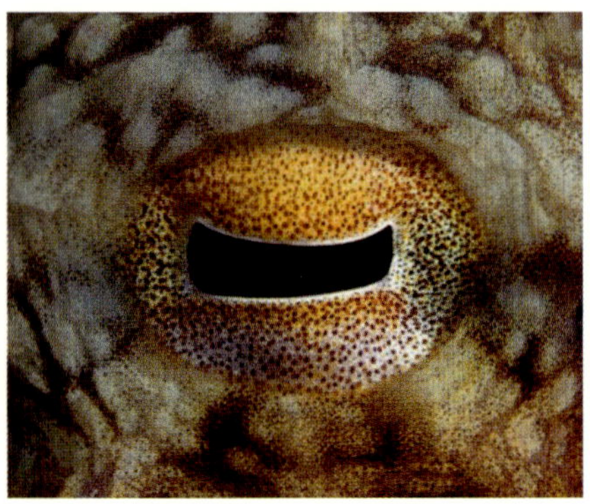

The Great Innovations

The egg

The answer to that eternal question "Which came first – the chicken or the egg?" is simple and without doubt. The egg certainly came first, long before chickens, mammals and even before all reptiles. It is presumed that early gastropods (marine snails) were the first to reproduce using eggs sometime during the Cambrian. As the eggs lacked a hard shell, there is no fossil evidence of them known.

The gastropod *Filifusus filamentosus* laying eggs.

Molluscs, fish and amphibians are amongst the animals that lay their eggs in water. Their eggs are able to exchange oxygen, carbon dioxide and waste products directly with the surrounding water. The ocean literally is the womb in which they develop.

The egg has remained the reproductive strategy of choice for thousands of marine species for more than 500 million years.

The twisted fibrous egg case of the Lightning whelk, a gastropod, often cast up on beaches.

Throughout their history molluscs have done it all in terms of reproductive strategies – individual eggs, anchored eggs, multiple embryos in egg capsules (with intra-capsular cannibalism), parasitic larvae, hard-shelled calcium eggs, even live birth and brooding. Gastropods evolved some of the most spectacular egg-laying strategies of any living organisms. While some eggs are piled somewhat randomly one on top of each other, many gastropods developed fibrous cases or gelatinous ribbons in which their eggs are protected and held to the substrate.

Jurassic ammonite eggs.
UK, Kimmeridgian 150 mya.

Transparent commensal shrimp *Urocaridella antonbruunii* clearly showing eggs contained within it. Comoros Islands, Indian Ocean.

Amongst the cephalopods there have been, and remain today, very different egg-laying strategies. Ammonites seem to have opted for large numbers of tiny eggs (with rare fossil evidence of these preserved), whereas *Nautilus* chose to lay small numbers of relatively large eggs (based on observations of the living species).

The egg has been a spectacular innovation that has remained the reproductive strategy of choice for thousands of species for more than 500 million years – especially in the marine world. The movement of life onto land, however, demanded further innovation.

Thousands of eggs of the Spanish Dancer nudibranch *Hexabranchus sanguineus* deposited in a rose-shaped gelatinous ribbon. Two-Mile Reef, Sodwana, South Africa.

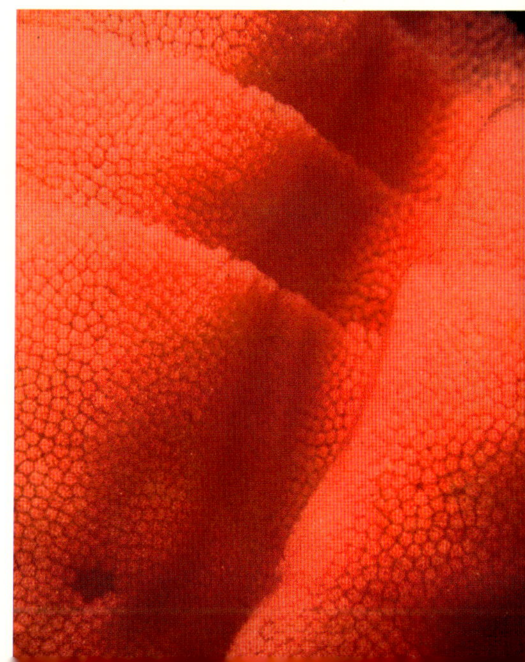

The Great Innovations

The amniotic egg

Laying eggs in water or on the ocean bottom is one thing, but to do the same on land required a different and more resilient approach. For animals to move from water to land, meant taking their own 'sea water womb' with them – within their bodily fluids.

Life on land involved a major redesign of every aspect of an organism, from breathing to reproduction. Vertebrates that lay their eggs on land, and those that retain the fertilized egg within the mother resulting in live births are called *amniotes* and include reptiles, birds, and mammals. The term *amnion* refers to the embryonic membrane produced during foetal development which serves to transport oxygen into the egg and expel carbon dioxide – it allows the eggs to breathe. Animals that lay eggs within which the foetus develops, such as most reptiles, are called *oviparous amniotes*. If the egg is retained inside the mother and develops there – such as in humans and ichthyosaurs – these are called *viviparous amniotes*. This characteristic evolved at least three times independently during the descent of placental mammals.

> *The amniotic egg was a critical evolutionary development that enabled life on dry land.*

Fishes and amphibians typically lay their eggs in water and together form a clade called the *anamniotes*. As the name suggests, anamniotes do not develop an *amnion* and are able to exchange oxygen, carbon dioxide and waste materials directly with the surrounding water.

The *amniotic egg* enabled animals to reproduce on dry land – free of the need to return to water for reproduction as is required of the amphibians. This critical evolutionary innovation enabled them to breed in drier habitats and lay bigger eggs and facilitated their spread around the globe into environments previously too hostile.

The amniotic egg went through a series of evolutionary steps. An early innovation was the gradual replacement of the gelatinous coating covering the amphibian egg with a fibrous shell membrane. This meant bigger eggs and improved gas exchange, permitting the embryo to reach full development before hatching.

Theropod dinosaurs typically laid oblong eggs. This almost complete nest showing pairs of eggs, each 180 mm long, laid from dual oviducts. Prepped in 3D, this image shows the bottom of the nest, while the top (seen in the reflection) revealed that most eggs are unhatched and contain embryonic material. *Macroolithus yaotunensis*, Late Cretaceous, 75–65 mya, Shixing, Nanxiong Basin, Guandong Province, Southern China.

Soft and flexible eggshells such as those laid by many reptiles, including snakes and turtles, are composed of an organic framework and loosely organized calcite crystals. These eggs collapse and shrivel after the animal hatches and are therefore unlikely to be preserved as fossils.

Further developments, like the hard, calcified egg shell, evolved later. Dinosaurs and birds lay eggs with shells made of calcium carbonate crystals which form a relatively thick structure of interlocking units. Fossilization is more likely to occur in rigid eggshells because they do not shrivel upon hatching.

Dinosaurs literally perfected the technology of hard-shelled eggs, as is evidenced by the many forms preserved in the fossil record. They passed these characteristics onto their direct avian descendants, the birds, whose eggs provide another clear evolutionary link to their therapod dinosaur ancestors.

Sauropods were the largest of all dinosaurs and typically laid round eggs. This fossil egg from a giant titanosaur is broken in half and tells the history of its fossilisation on the banks of a river. The eggs were covered in mud during a flood and gradually filled with water. The minerals present in the water crystallised from the edge of the egg to the centre, creating this fabulous concentric agate formation. *Megaloolithus* sp., Cretaceous, approx. 70 mya, Patagonia, Argentina. Size 200 mm.

The oldest dinosaur eggs known. They are 190 million years old from an early sauropod dinosaur *Massospondylus*. They were found in Golden Gate Highlands National Park in South Africa in 1976, and they contain embryonic dinosaur bones, with one apparently entombed in the act of breaking through the eggshell – see close-up below.

The Great Innovations

Life moves onto land

Tiktaalik is an extinct genus of lobe-finned fish and one of the first individuals that moved onto land. The first well-preserved *Tiktaalik* fossils were discovered in 2004 on Ellesmere Island in Nunavut, northern Canada by Ted Daeschler and Neil Shubin.

Tiktaalik lived in the Late Devonian period, about 375 million years ago, and is thought by many to represent the evolutionary transition from fish to amphibians. It shows adaptations to the oxygen-poor shallow-water environment at the time, which may have played a part in its evolution onto land. It already displayed many features of four-legged animals (tetrapods) including front fins with arm-like skeletal structures similar to a crocodile, a shoulder, elbow, and wrist.

Tiktaalik *is the most important fossil since Darwin*

Nature

Tiktaalik reconstruction.

Tiktaalik has been called a transitional fossil, as it is to tetrapods what *Archaeopteryx* is to birds. That unique position is already being contested by discoverers of trackways, in Poland's Holy Cross Mountains, which may have been made by early tetrapods around 427 mya – fifty million years before *Tiktaalik*. It may or may not be the common ancestor of all vertebrate terrestrial animals, but it still remains a strong contender.

The Great Innovations

Lest we forget the plants

Life on land would have been unthinkable without plants. Green algae from the oceans likely fuelled the development of primitive land-based plants in the Ordovician, ferns during the Silurian, gymnosperms (primitive plants with naked seeds such as pine cones) during the Carboniferous and eventually angiosperms (flowering plants) during the Jurassic.

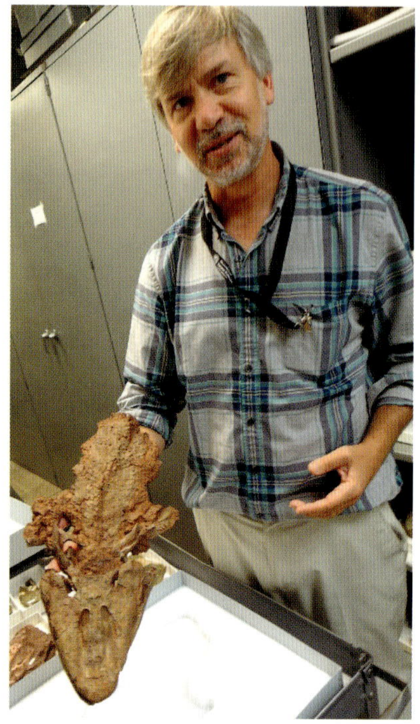

Ted Daeschler with a fossil Tiktaalik during our meeting at the Academy of Natural Sciences in Philadelphia.

We mustn't forget the plants, without whose prior invasion of the land none of the other migrations could have happened.

Richard Dawkins in *The Ancestor's Tale*

THE OCEANIC EXTINCTIONS

Placoderms, trilobites, belemnites, ammonites & sea dragons

Not all of the oceanic extinctions were tied directly to mass extinction events. But, it would be somewhat arrogant and anthropocentric to call these departures 'minor'. If you're a trilobite then your extinction is certainly not a trivial matter. However, in the scheme of all life on earth, the extinction of trilobites has certainly proved to be of minor importance.

Oceanic Extinctions

The Placoderms

Among the first jawed fish in the Silurian 420 million years ago, there emerged a spectacular group that unfortunately didn't make it for too long – just 70 million years, a geological blink of the eye.

Placoderms were the first fish to develop pelvic fins, the precursor to hind limbs in tetrapods, as well as true teeth. They were covered by articulated armoured plates and the rest of the body was scaled or naked, depending on the species. *Dunkleosteus* is the largest placoderm that ever lived, growing up to six meters in length and to more than one thousand kilograms in weight – the same as a large modern shark. It could quickly open and close its jaw, like modern day suction feeders such as the large rock cods.

Their armoured frontal sections easily fossilised, but the back parts of *Dunkleosteus* are mostly unknown. Placoderms succumbed to the mass extinction at the end of the Devonian 360 million years ago.

Fossil head of *Dunkleosteus*
1.7 m in length.

The spiny trilobite fossils above and right are from the famous Vilpovitsy quarry, St. Petersburg, Russia. Ordovician 460 mya.

Above: *Paraceraurus exsul.*

Below: *One of the earliest* trilobites *Eccaparadoxides* sp. from Morocco. Middle Cambrian 520 mya.

Oceanic Extinctions
Trilobites

A magnifenctly ornamented trilobite *Boedaspis ensifer.*

The Cambrian Period brought about the first eyes, antennae and claws, plus evolved nervous systems that allowed for interaction, scavenging and predation. The trilobites were notable pioneers of these evolutionary innovations and became the most diverse and abundant of the Cambrian animals.

They had a hard exoskeleton typical of arthropods, composed of calcite and calcium minerals in a lattice of chitin covering the upper surface and curled round the lower edge. Their great success was helped by their multifaceted eyes. In some species their bodies were adorned with the most spectacular spines which have been exposed for the first time by improved fossil preparation techniques during the last few decades.

The earliest Cambrian trilobite fossils are about 530 million years old, but they were already diverse and geographically dispersed, suggesting that they had been around for quite some time before that. They serve well as stratigraphic marker fossils for dating the Cambrian explosion of life. As the first truly mobile inhabitants of the seabed, they dominated it for more than 200 million years. They are often described as "the first arthropods" and as "the ancestors of today's insects, spiders, lobsters, crabs, shrimps and scorpions", but the conclusion today is that these groups evolved separately from common worm-like ancestors in the early Cambrian.

Trilobites began a gradual decline in the Devonian around 400 million years ago and disappeared totally in the mass extinction at the end of the Permian about 250 million years ago. There are more than 10,000 fossil species recorded. Their similarity to living arthropods has led to the assumption that trilobites must also have produced eggs. In January 2017 a new report proclaimed the "First-ever discovered trilobite eggs paired with a fossil trilobite" apparently confirming that this was indeed the case.

Oceanic Extinctions
Belemnites

Belemnitida is an order of cephalopods that thrived during the Jurassic and Cretaceous periods, and they, and their bullet-shaped fossil rostrums, are commonly referred to as belemnites.

Belemnites were externally squid-like in appearance but, unlike modern squid, had a chambered phragmocone (a shell structure typical of cephalopods such as Nautilus and ammonites) made of aragonite and with the calcite rostrum at its tip. Because calcite is geologically more stable, the bullet-shaped rostrum is often the only part of the animal preserved, often in very large numbers. In the suborder Belemnotheutina, the rostrum was only present as a thin layer.

Belemnites possess ten arms each equipped with 30–50 pairs of inward-curving hooks used for grasping and holding prey. They were consumed in large numbers by sharks, ichthyosaurs and other marine reptiles. Fossil belemnite hooks are often found in their stomachs. They were in fact more closely related to modern cuttlefish, which also have an internal shell, than to squid. Belemnites lacked the pair of specialized tentacles present in modern squid.

Belemnites were among the early coleoids to evolve. They thrived for around 140 million years until they went extinct at the end of the Cretaceous.

Rare fossils in which the soft parts are preserved show that, like modern coleoids, they possessed an ink sac, hard beaks and large eyes.

This rare belemnite has been identified as belonging to the new genus *Clarkeiteuthis montefiorei* (Buckman 1880) described by Fuchs, Donovan & Keupp in 2013. Desmond Donovan commented "This specimen is what has been long time called *Phragmoteuthis conocauda* or *Belemnnoteuthis montefiorei*... even *Phragmoteuthis huxleyi*. No one has been quite sure till now". This specimen has rare 3D preservation with the top preserved in shiny nacre, and some hooks persevered. The phragmacone has been fully preserved but folded underneath. The combined two images of the top and bottom of the fossil nodule show off a unique composite view of the animal.

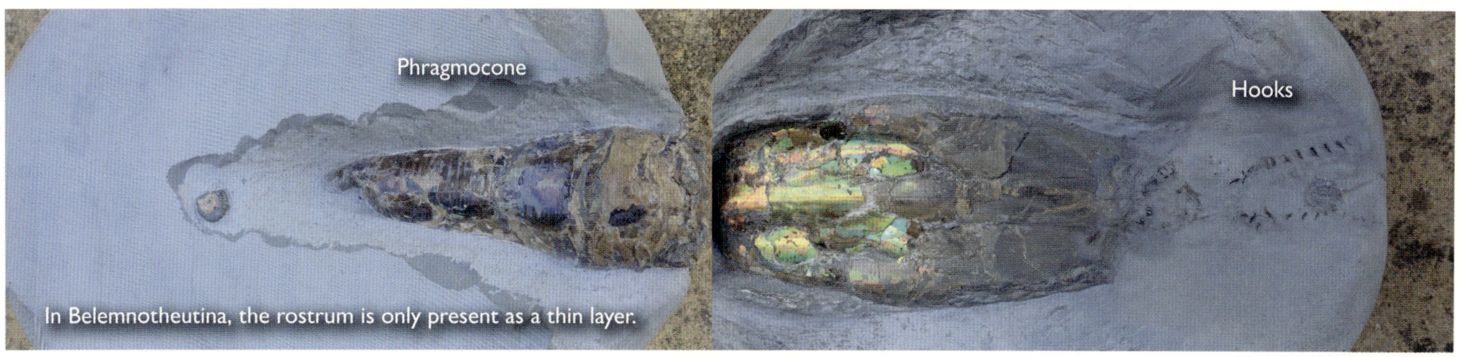

In Belemnotheutina, the rostrum is only present as a thin layer.

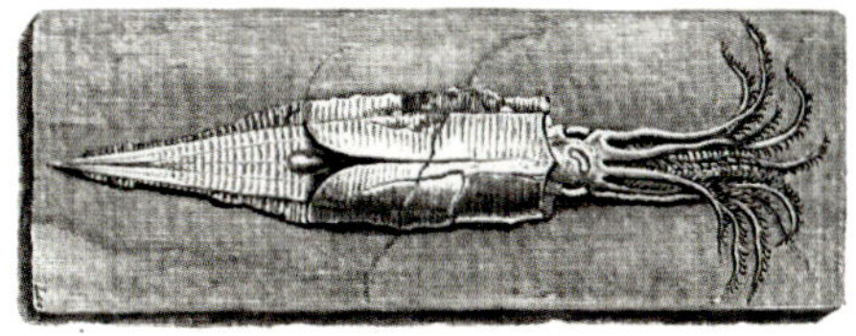

Belemnotheutis antiquus as illustrated by S.P. Woodward for Gideon Mantell's Petrifactions and Their Teachings, 1851.

A matrix made up of hundreds of fossil belemnite restrums.

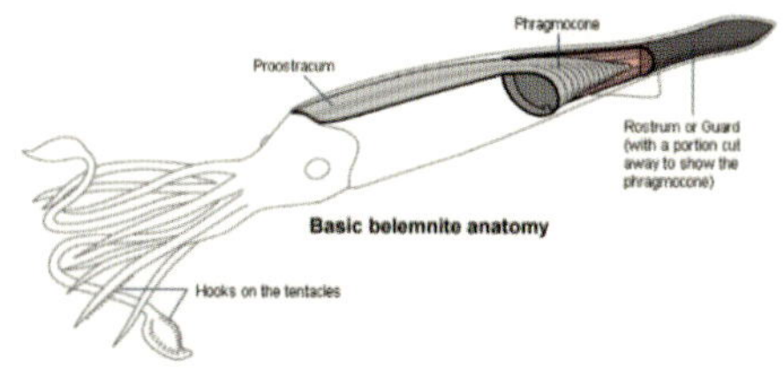

A belemnite rostrum preserved in opal, Coober Pedy, Australia. Cretaceous.

Imagine the ancient Cambrian ocean, in which all animal life is restricted to the sea floor and even rooted to it. Small molluscs and trilobites graze on algal mats amongst primitive echinoids. Suddenly (in geological time) some primitive molluscs develop a chambered shell that gives them the ability to rise off the ocean floor and move into the water column. From this new perspective they survey the playing field in a whole new way. Imagine the predatory advantage this provided them and their kin. Only the freely floating jellyfish had managed this any earlier.

Modern diver inserted here for scale only.

Oceanic Extinctions
Ammonites

A mass mortality of chambered nautiloid shells (Orthocerida) from the Silurian strata of Tafilalt, Morocco.

Shells became increasingly coiled.

Many of the early shelled cephalopods, including *Plectronoceras* and the Orthocerida, had shells that were almost straight. Over time, some species coiled their shells until they began to look more like those familiar to us as nautiluses and ammonites. The straight shell shape finally disappeared sometime during the Triassic.

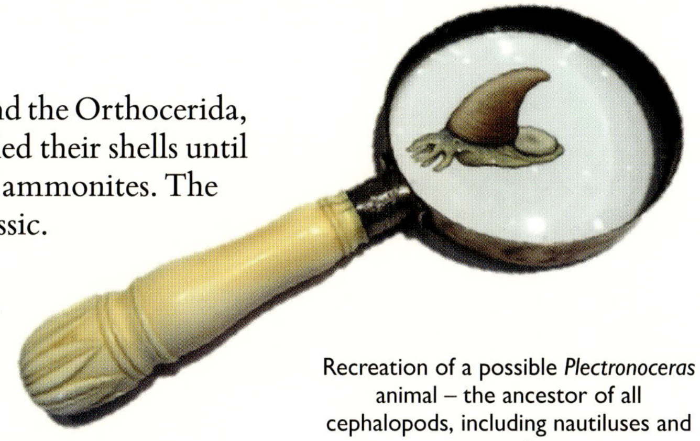

Recreation of a possible *Plectronoceras* animal – the ancestor of all cephalopods, including nautiluses and ammonites.

Nautiloids were the dominant inhabitants of the water column until the arrival of the first jawless fishes – they were dominant in the open water for tens of million years. Without serious competitors they populated the ocean in vast schools and some grew to enormous sizes. One genus, *Cameroceras*, grew up to 10 meters long and they were certainly the earliest large animals in the ocean.

Sometime in the Devonian an important new cephalopod group evolved: the ammonites. Their shells were outwardly similar to the coiled nautiloids, but internally their chamber walls developed into much more complex shapes. The tube connecting the various chambers – the siphuncle – was placed against the outer shell wall rather than centrally.

For almost all of the next 200 million years the ammonites would represent one of the largest and most diverse animal groups in the ocean – see chart on page 102. Because of their abundance and extreme diversification, ammonites are excellent index fossils allowing the age of rocks to be determined by the ammonites found within them.

New species appeared, evolved and disappeared with regularity – a phenomenon that was especially true of heteromorph ammonites. Ammonites appear to be excellent examples of *sudden evolution*, the reasons for which are not yet understood. *Suddenly* in geological terms could still mean intervals of thousands or millions of years.

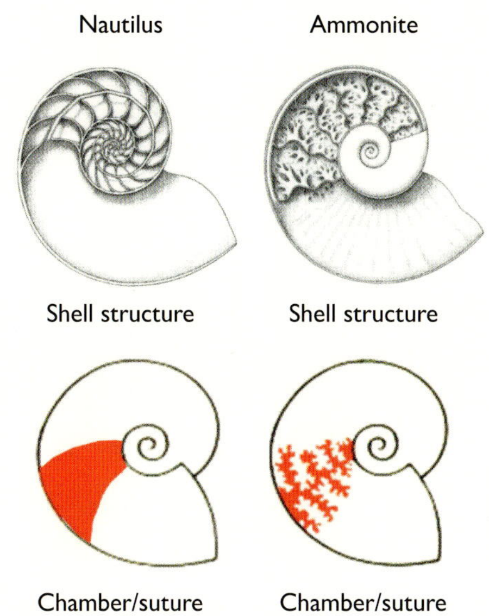

Several thousand fossil ammonites preserved In a septarian nodule. This mass mortality is mainly composed of *Promicroceras planicosta* and juvenile *Xipheroceras ziphus* ammonites, these probably representing, respectively, the male (microconch) and female (macroconch) partners of a single biospecies. The cause of this mass mortality is unkown, but may have been due to anoxic conditions (perhaps caused by an algal bloom, or a storm event disturbing stagnant sea floor sediments) selectively killing small ammonites. It is interesting that this nodule contains three distinct layers of ammonites, perhaps deposited in subsequent years and then fossilised within this discoidal, calcareous 'mud ball'. After fossilisation, the ball cracked, and the cavities were filled with mineral-rich water in which calcite crystals formed – creating the veins of crystal that are evident on both sides of the nodule.

This is the rear side of the same septarian nodule polished to a high gloss. It shows off the cross-sections of the *Promicroceras* and *Xipheroceras* ammonites, revealing a unique view of their internal structure and the calcite crystal veins. There is also an imprint of a larger ammonite *Asteroceras blakei* at top left. This large nodule is from the Jurassic (Sinemurian 190 mya) and was found at Marston Magna, Somerset, UK.

Size 640 x 500 mm.

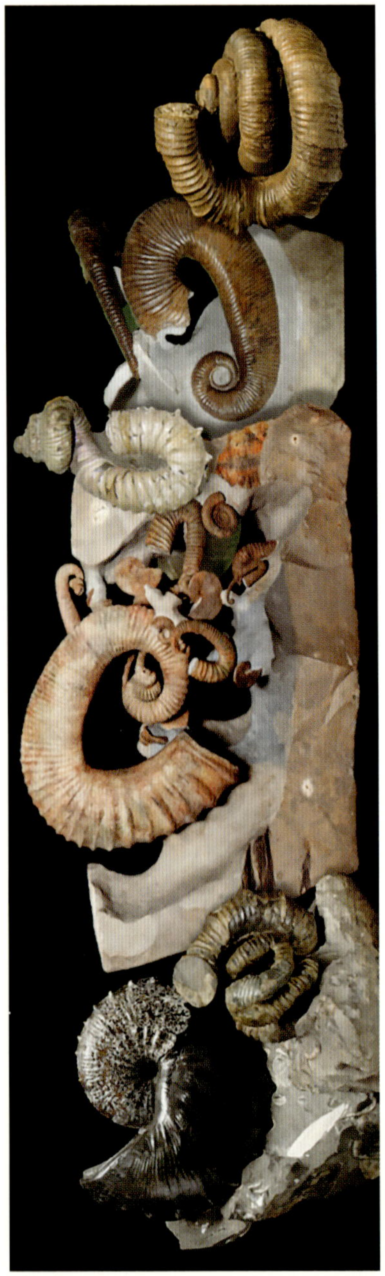

Heteromorph ammonites

Three times during the long reign of the ammonites, certain species uncoiled their shells into the most bizarre shapes. These so-called *heteromorph* ammonites are grouped into the sub-order Ancyloceratina shown in three pink areas in the ammonite family tree at right.

These three *heteromorph revolutions* initially attracted limited focus and were thought about as nothing more than evolutionary oddities. Late in the 20th century, however new fossil preparation tools allowed their delicate shapes to be properly exposed. They turned out to represent more than half of ammonite diversity in the middle of the Cretaceous Period.

Some heteromorph ammonites, such as the *Diplomoceras* shown below, were extremely large. This shell is 1.6 m long with commensurately large body chamber. The last chamber wall is shown by the red arrow indicating that the body chamber is around 90% of the total shell volume. Such large body chambers are unusual in the ammonite world where they would typically represent no more than 50% of total shell volume. They must have housed large animals that likely had to be voracious predators to feed such a large body mass.

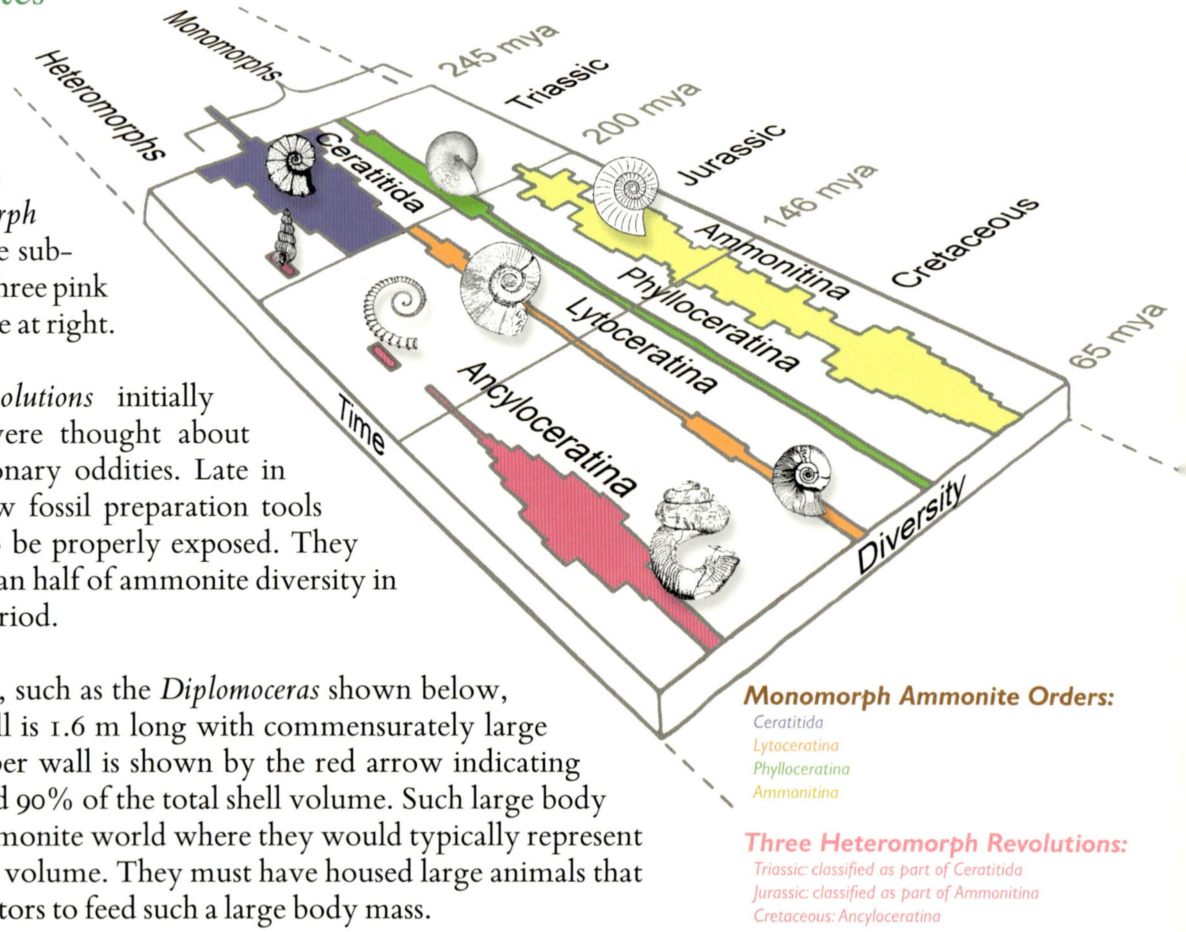

Relative diversity = approximate number of families.
Based on Patrice Lebrun Ammonites *March 2008.*

An animal to fill this volume could have weighed in at around 20 kg and its tentacles might have extended more than two meters outside the shell. Because only small parts of ammonite soft parts have been found preserved as fossils, we can only guess at the shape of this large animal. Did it have long octopus-like tentacles, or did it perhaps have an enormous umbrella-like web between its arms to enable it to snare large amounts of plankton to sustain it?

With this variety of shell shapes and possible lifestyles, it does seem somewhat strange that, during their last five million years on this planet, some heteromorph ammonites should end up with straight shells, notably the prolific *Baculites* genus, this took their shell shape almost back to where it all began with their straight nautiloid *Orthoceras* ancestors.

By the end of the Cretaceous Period 66 million years ago, all ammonites were extinct. They had thrived for more than 180 million years. The nautilids survived the extinction of the ammonites apparently by moving into ever-deeper water avoiding the acidification of the seas, and by altering their diet to the extreme environment at depth. Today *Nautilus* are pure scavengers, never eating live food.

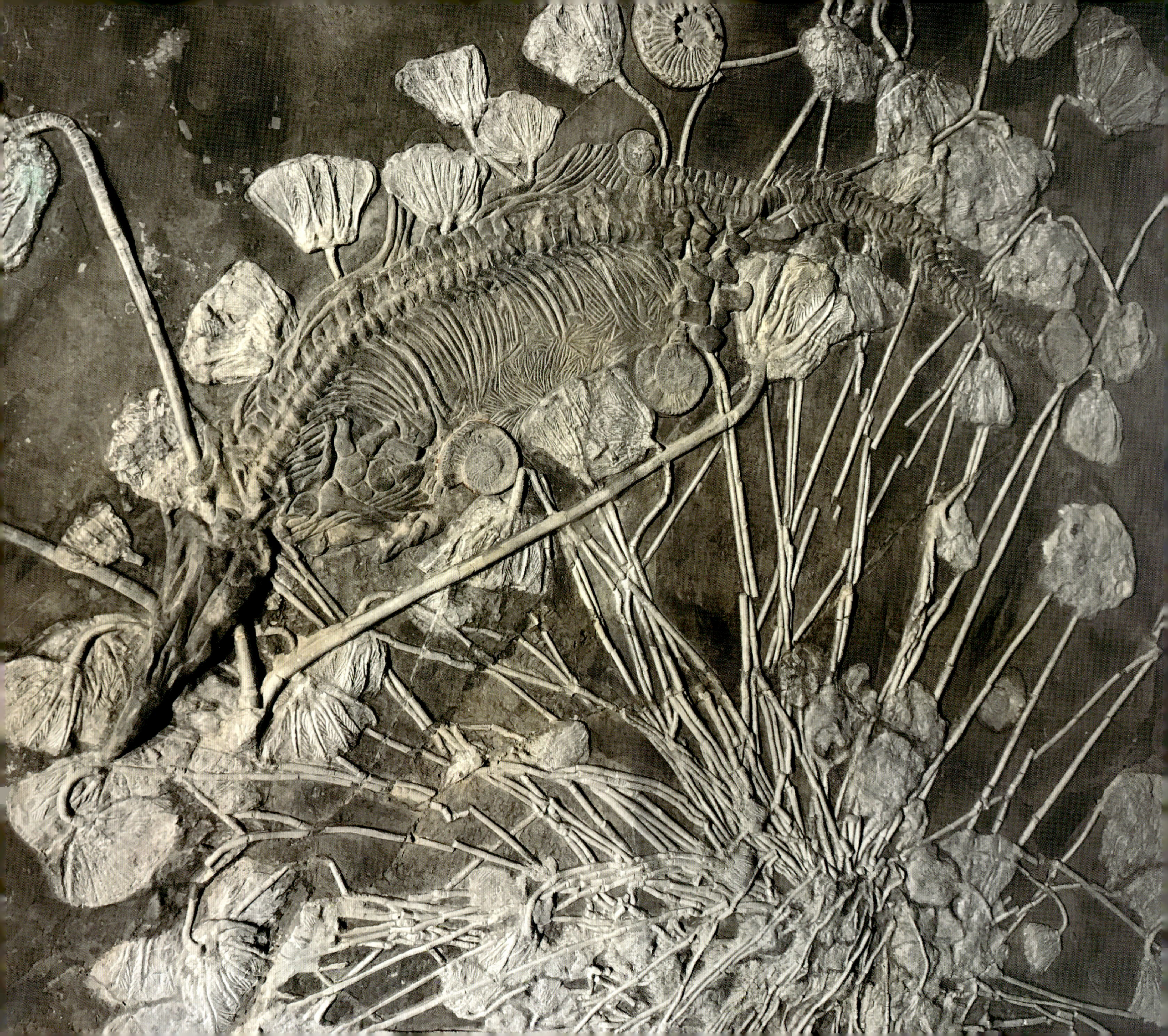

Oceanic Extinctions
The Sea Dragons

An early Sea Dragon, a *Thalattosaurus* fossil found at Xingyi, Guizhou Province, China. Triassic 230 mya, size aprrox. 2 m, amongst *Traumatocrinus* crinoids and ammonites.

Despite their spectacular success on land, some reptiles returned to the ocean, eventually becoming the top predators in their environment.

Thalattosaurus and *Keichousaurus* are two good examples of such early Sea Dragons that developed from terrestrial ancestors soon after the end-Permian extinction and flourished during the Triassic.

Thalattosaurus was a shellfish-eating reptile with paddle-like limbs that grew to around 2 metres in length. Many fossils of this genus are currently being unearthed in China. Fossils of the small marine reptile *Keichousaurus* from China have been popular with collectors for a few decades now.

'Sea Dragons' is a common name for the iconic marine reptiles of the Triassic, Jurassic and Cretaceous seas. The term was attributed to them by Thomas Hawkins in the early 19th century as their fossils emerged into public consciousness.

This gang of Sea Dragons was to be joined during the Early Triassic by the much larger ichthyosaurs.

'Plesiosaurs' is the generic term used to encompass both the typically long-necked plesiosaurs, and the short-necked pliosaurs, as the one shown above. The pliosaur *Kronosaurus* grew to around 10 m.

All of early Sea Dragons became extinct at the end of the Triassic, except for the ichthyosaurs and the long-necked plesiosaurs, which may have evolved from a shared ancestor. The survivors continued to diversify till the end of the Cretaceous and their descendants included the elongated elasmosaurs and the giant short-necked pliosaurs.

The first fossils of the plesiosaurs, collected by Mary Anning at Lyme Regis in the UK in the early 19th century, were so spectacular and unusual that they were initially dismissed as fakes by Georges Cuvier. Their basic body plan is also said to have inspired the profusion of tales about the Loch Ness Monster in Victorian times.

Keichousaurus hui fossil at the North American Museum of Ancient Life. Size 200 mm.

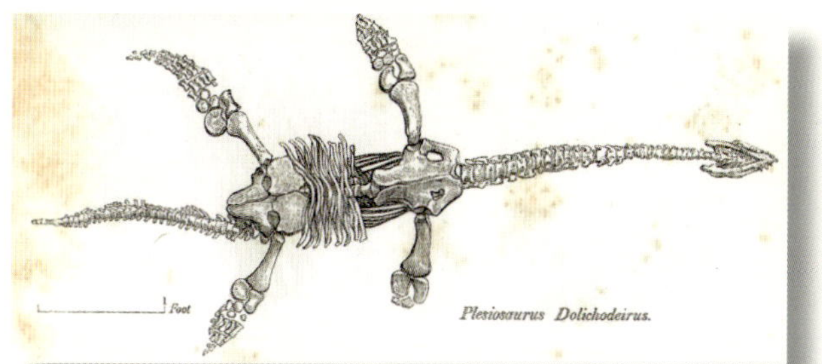

Engraving of a fossil plesiosaur Thalassiodracon hawkinsi, which was found in Somerset, UK, in the early 19th century. Size approx. 1.7 m.

Mary Anning was one of the first to unearth fossils of the fabled Sea Dragons. These are of some of her finds, an ichthyosaur (now regarded as a specimen of *Temnodontosaurus platyodon*) and a plesiosaur *Plesiosaurus dolichodeirus* – shown is her original sketch (at right) plus notes about her discovery of the first plesiosaur in 1824. The original fossil specimens of both are at the Natural History Museum, London.

A Drawing of the **ICHTHYOSAURUS** *found at* LYME,

In the Collection of

THOMAS HAWKINS ESQr F.G.S.

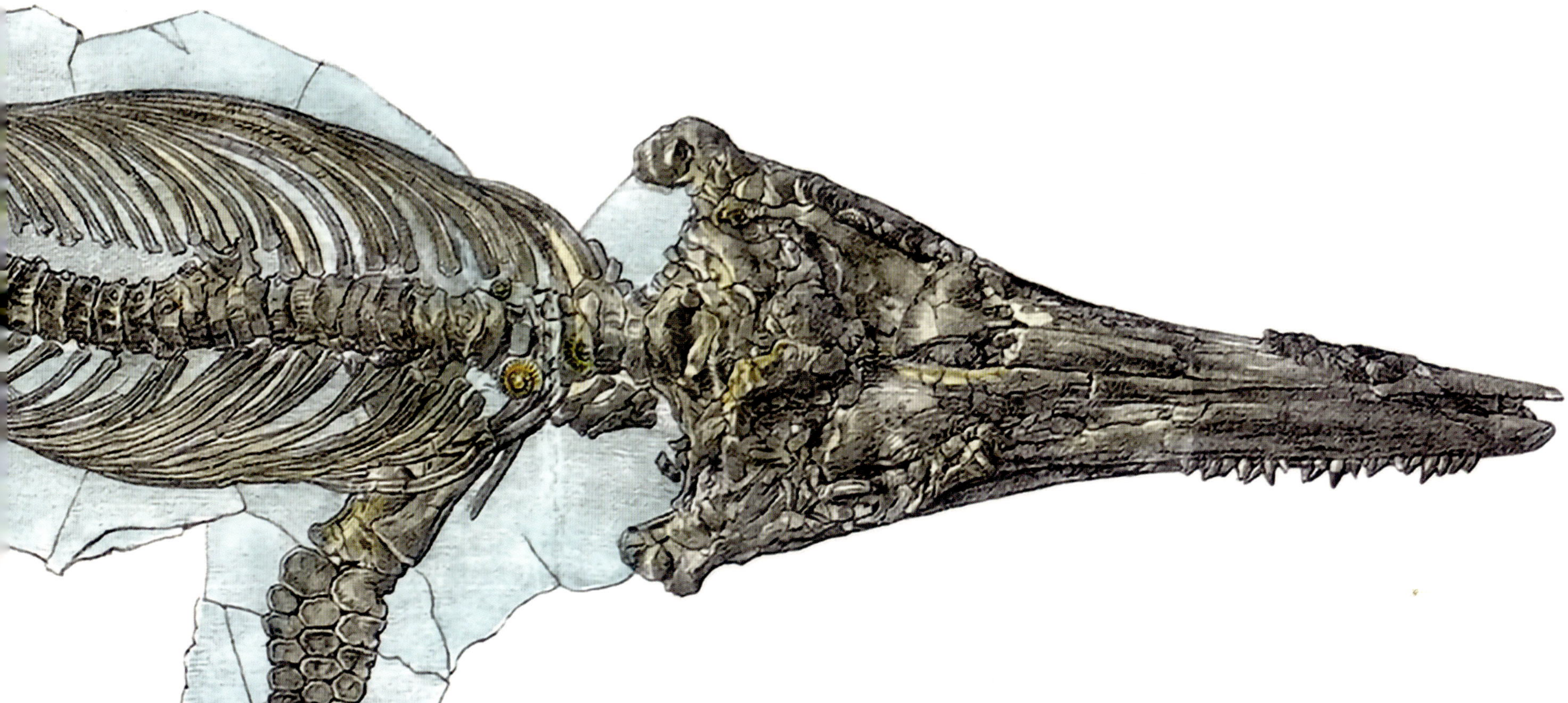

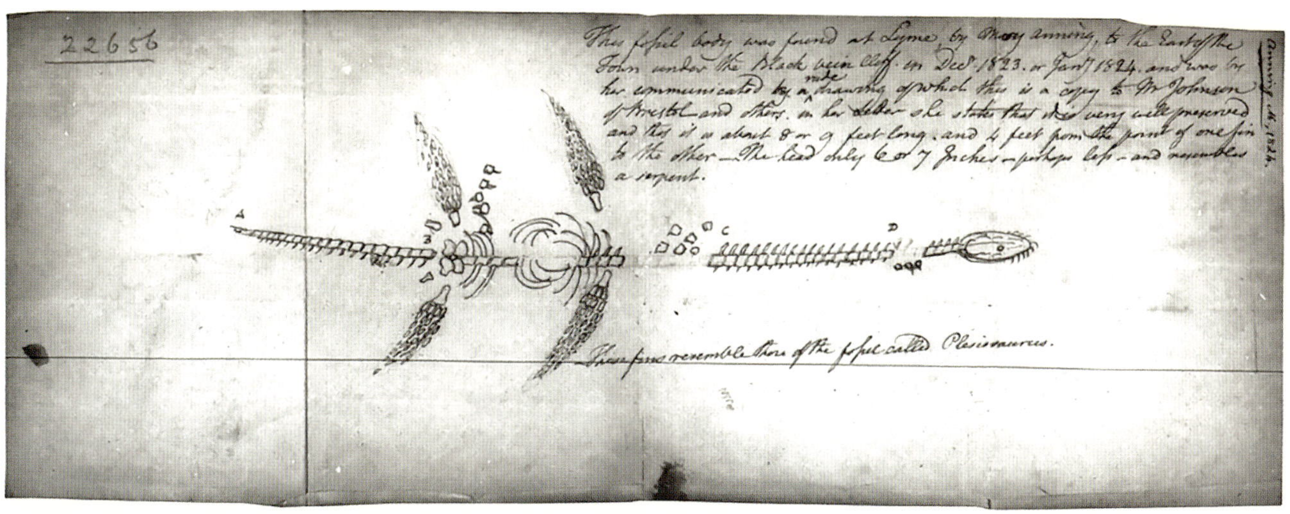

Ichthyosaurs were the first reptiles to evolve fish-shaped bodies and to reach truly gigantic proportions.

Ichthyosaurs were the first tetrapods to evolve fish-shaped bodies, in the Triassic. Other Sea Dragons, including the Cretaceous mosasaurs and more recently whales, were to end up with similar streamlined bodies, but developed independently and from totally different ancestors. The exact evolutionary steps and the environmental pressures that may have led to these profound changes in lifestyle are still very poorly understood and much debated.

Ichthyosaurs were the first animals to attain truly gigantic sizes, with the largest complete fossil specimens reaching more than twenty metres in length. The largest is from the Late Triassic British Columbia, measuring 21 m in length, . Some partial fossil finds, such as chunks of giant jaw bones found in the UK, indicate possible sizes over 30 m.

What is clear is that while dinosaurs roamed the land, various marine-saurs evolved to thrive as the top predators in the ocean, enjoying the abundant food sources available in Jurassic and Cretaceous seas. Some ichthyosaurs have been found with their last meals preserved. Most often they contain squid hooks, but also fish scales. The rarest stomach contents found so far include hatchling-sized turtles and a bird.

Ichthyosaurs had amongst the largest eyes in the animal kingdom, and this specimen, below, is one of the largest Cretaceous ichthyosaur skulls yet found – more than 2 m long, with teeth each more than 65 mm long. From the Albian black shales of Provence, France. About 100 mya. The last ichthyosaur fossils known date to about 90 mya.

Marine reptiles in pursuit of Jurassic squid.

A rare fossil ichthyosaur in my collection is from the Holzmaden quarries in Germany has been dubbed the *Kamikaze Ichthyosaur*. Its mouth has been crushed and several vertebrae have popped out with one lodged firmly in its eye. Its stomach contains many fossilised squid hooks.

There is some controversy as to how exactly this and similar finds fossilised. Some say that the damage to its skull occurred after death, when it dropped into the mud "head first". Alternatively the breaks and displacement could have happened after fossilisation. My own interpretation, based on my diving experience with live marine predators, is that the ichthyosaur could have been feeding by 'flying blind' through a huge school of squid – until it hit a proverbial 'brick wall', dropped dead onto an anaerobic seafloor and was preserved with its broken bones, before any predation could take place.

The Kamikaze Ichthyosaur, Jurassic, Holzmaden, Germany.

Reconstruction of Jurassic squid showing arms with hard hooks.

This spectacular fossil of a giant long-necked plesiosaur has been attributed to the elasmosaurs as *Zarafasaura oceanis*. This is one of the latest discoveries in Morocco, from the Ouled Abdoun Basin phosphate deposits, near Khouribga.
Length 8.4 m. 75% complete.
Cretaceous, Maastrichtian 72–66 mya.

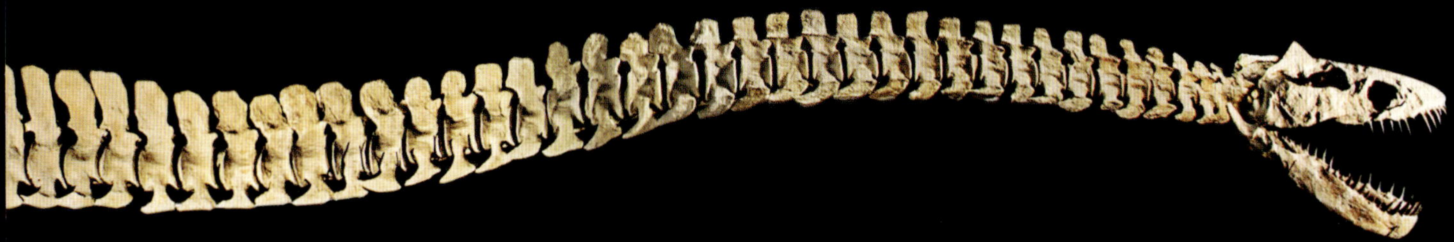

Compare this to one of the first plesiosaurs found in the early 19th century. This Jurassic specimen was less than 2 m long.

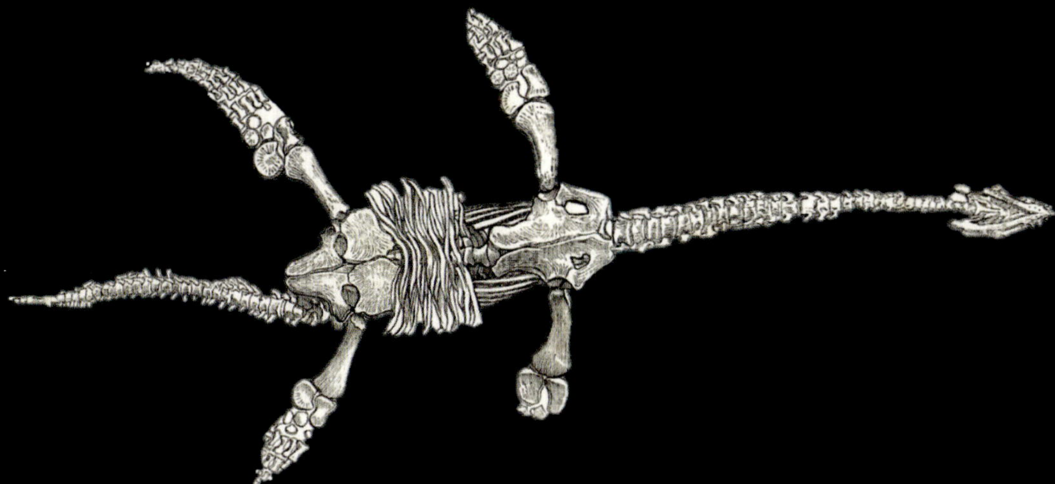

Mosasaurs

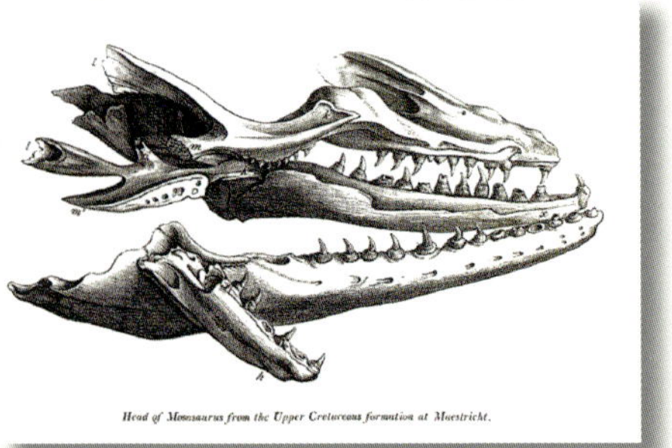

The first mosasaur skull.

Mosasaurs likely evolved from terrestrial lizards in the Early Cretaceous, quite independently from ichthyosaurs and their relatives. With the extinction of the ichthyosaurs and pliosaurs during the last 25 million years of the Cretaceous, mosasaurs became the dominant predators in the ocean. They grew up to 17 m in length.

Discovery of the first mosasaur at Maastricht.

One of the first mosasaur skulls was discovered in an underground limestone quarry at Maastricht, Holland, in 1764. It was acquired by Dr Hoffman, a local surgeon, but ownership was subsequently contested by the landowners. The remains were sent to Paris and in 1808, Baron Cuvier published the first comprehensive description of a mosasaur and proposed that its ancestry lay somewhere between iguanas and lizards. In 1822, the specimen finally received its scientific name when Conybeare named it *Mosasaurus*. Some years later Gideon Mantell added the species name *hoffmanni* in honour of Dr Hoffmann. It is now believed that Cuvier may in fact have been working with a copy of the original skull and that Dr. Hoffman may have altered some of the original find to make it more attractive.

Monitor lizard *Saniwa ensidens* from the Green River Formation in Wyoming ... it is also the largest at 190cm in length. Eocene 50 mya. They are believed to have evolved from the same ancestral line that gave rise to mosasaurs and the Komodo dragon.

In the past century many complete fossil mosasaurs have been found in the phosphate mines of Morocco (see image on the facing page) and many have been assembled from partial finds. Their spectacular size and teeth were undoubtedly an inspiration for the movie *Jurassic World* – even though its size was greatly exaggerated, and they never existed in the Jurassic.

By the time of the mass extinction during which the mosasaurs also met their end, the ichthyosaurs had already been extinct for 25 million years, due to massive climate change during the Cenomanian 90 mya.

Some fans of the Jurassic marine reptiles denigrate the Cretaceous mosasaurs by calling them "nothing but overgrown seagoing lizards, vastly inferior to the larger ichthyosaurs and pliosaurs". But evolution had other ideas, and in the Cretaceous, mosasaurs might well just have outsmarted and outsized their older competitors.

For all these Sea Dragons there was a major obstacle to making a complete return to the water – the reptilian egg was suited to terrestrial life but its porous leathery shell would have allowed water to enter, drowning the embryo. This legacy still binds turtles to the shore for reproduction. From pregnant ichthyosaurs found in the earliest Triassic of China and some Jurassic ichthyosaurs from Germany fossilised in the act of giving birth, we know that they evolved the ability to bear live young. The plesiosaurs were long thought to have hauled themselves ashore to lay eggs. But in 2011 a paper in Science reported the first plesiosaur fossil found with an embryo preserved, showing that plesiosaurs too were in fact viviparous. It is highly likely that the Sea Dragons precursors already bore live young while still living on land.

At any rate, by the end of the Cretaceous, all the major predators at the top of the food chain, the so-called Sea Dragons, were gone forever. Sad, really. But what an opportunity for new life to evolve at the top of the food chain in the ocean – and move into the space left by the Sea Dragons.

BEHOLD THE MISBEGOTTEN HELICOS
Lebedev 2009
Todd Marshall 2002
Karen Carr 2012
Todd Marshall 2005
Mike Williams + Kathy Elbaum 1973 (with tongue firmly in cheek)
IVY RUTZKY 1998
Russell Wodehouse Pizza Cutter 1995
Ray Troll 1993
Purdy + Parrish 2008
Ray Troll + Gary Staab 2001
Ray Troll 1997
Alex Broadbent 2012
Karpinsky 1911
Karpinsky 1899
John Long 1995
Richard Ellis 2001
Karpinsky 1902
Van Den Berg 1953
Eaton 1962
lower jaw
Heinrich Harder 1906
Svend Erik Bendix Almgreen 1966
Obruchev 1953
Lebedev 2009

Oceanic Extinctions

The weirdest and fiercest sharks

Sharks and rays are part of the group Chondrichthyes. What differentiates them from bony fish is their cartilaginous skeleton. That difference is a real problem for the fossil record – cartilage, the stuff your ears and nose are made of, is softer than bone, and does not fossilise easily. Consequently, there are very few complete bodies in the shark fossil record, which is mostly made up of plenty of teeth – clear evidence that, then as now, sharks naturally shed thousands of teeth during their lives and that these fossilise relatively easily and often.

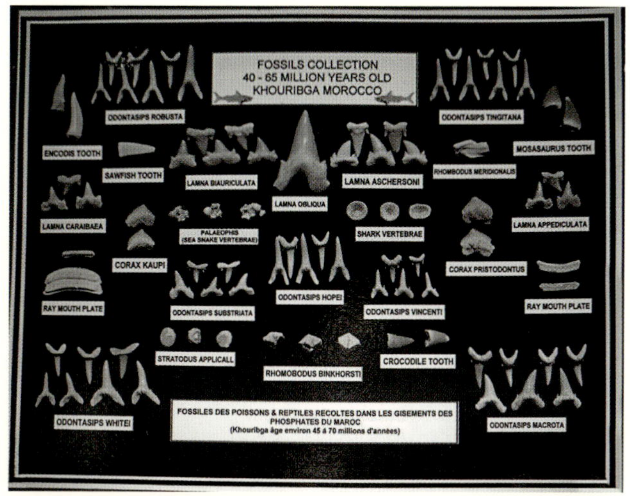

Fossilised shark teeth are known from the early Devonian onwards and have become one of the most popular fossils for kids and adults – exactly because they are so prolific in rivers and on beaches and, once you know where to look, you can often find hundreds. Because shark teeth are so distinctive, you are mostly able to tell the species from its teeth. Even an amateur collection is soon able to tell tales of the ancient life of sharks – who they were and what they ate.

The dangers of fragmentary evidence

The further back you go in shark prehistory, however, the more bizarre the conclusions drawn from fragmentary evidence become. One such example is *Helicoprion*, a widespread genus of shark that lived from about 290 to 270 million years ago. All that has been found since the first find in 1899 are parts of tantalising whorls of teeth. The debate continues today about what animal could possibly have surrounded these teeth. The dilemma of competing theories is brilliantly illustrated in Ray Troll's terrific summary of the various theories on the facing page.

The extinct *Carcharocles megalodon* shark was amongst the most fearsome that ever lived. It had individual teeth as big as your hand, and by comparison with present-day sharks, we can estimate that it had a mouth you could stand up in. Its giant fossilised teeth are frequently found and enthusiastically collected – with tough competition for any tooth over six inches in size.

The megalodon shark is traditionally believed to have lived between 23 and 2.6 million years ago but we recently found some vertebra and half of a six-inch tooth (pictures at bottom of page) in southern Africa in what appeared to be much earlier deposits. Analysis of nannofossils (the fossils of minute planktonic organisms, including unicellular algae) in the surrounding matrix, done by Jackie Lees at University College London, confirmed these to be from Late Maastrichtian deposits, specifically nannofossil subzone UC20aTP. (Dr Jackie Lees personal comment, 2018; University College London).

Example of a nannofossil from the 'megalodon' matrix.

It seems that if not a megalodon, then a very similar shark must have lived in the ocean at least 66 million years ago, 43 years before the presumed emergence of *Carcharocles megalodon*. Of course the megalodon experts can't quite believe it. Isn't science and discovery fun!?

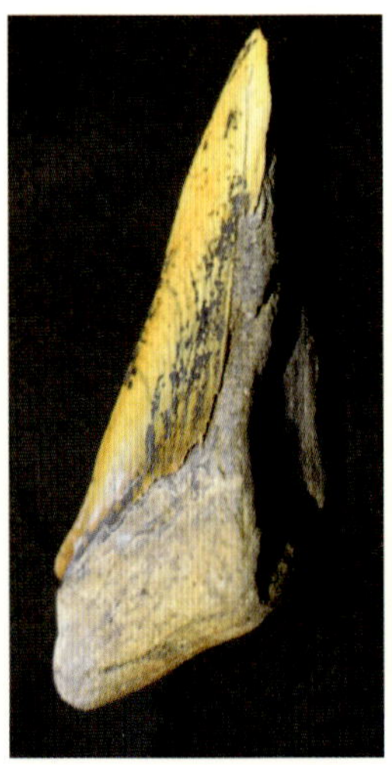

Evolution and extinction continued

Evolution is mostly spoken about in the past tense – and often assigned to the deep distant past. Of course, we know that evolution never ends, and that *Homo sapiens* may be playing an ever-stronger role in it.

Science is born from an act of humility: not trusting blindly in our past knowledge and our intuition. We learn nothing if we think we already know.

Carlo Rovelli,
Reality is not what it seems 2016

With the relentless march of science and technology, we are making evolution different, faster, and more creative – sometimes even while trying to arrest change.

Evolution and extinction will of course continue relentlessly into the future, with or without us, even as we influence both evolution (by selectively breeding our favourite animals) and extinction (by pursuing our preferred lifestyles).

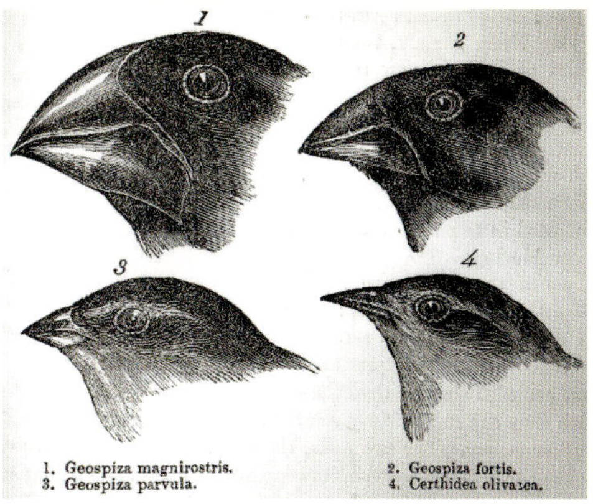

Traditional views suggested that nature does not make a jump, and Darwin assured us that all change is gradual over long periods of time. However, his report on the beaks of Galapagos finches in his 1839 *Journal of Researches*, now known as *Voyage of the Beagle*, already hinted at their fairly quick evolution into 13 distinct species based on different food sources from nuts to seeds, and fruit, flowers to insects. From the fossil record we have seen that evolutionary change can be unpredictable and sudden, and likely not only in response to environmental change. New life forms may search for suitable environments rather than simply adapting to their current ones. Evolution is highly opportunistic – it moves to fill available niches.

We should realise that extinction is vital to the development and renewal of life, no matter how difficult that is to accept when it gets *personal*. We would do well to remember that the 'biggest' and 'best' are often the most vulnerable to extinction – could we perhaps be in that category too?

Extinction is vital to the development, renewal and continuity of life.

But let's move on from evolution and extinction, and focus on celebrating the results of the unique 500-milion-year experiment that we have explored in this chapter – let's celebrate the amazing life in today's ocean, its incredible resilience and spectacular continuity.

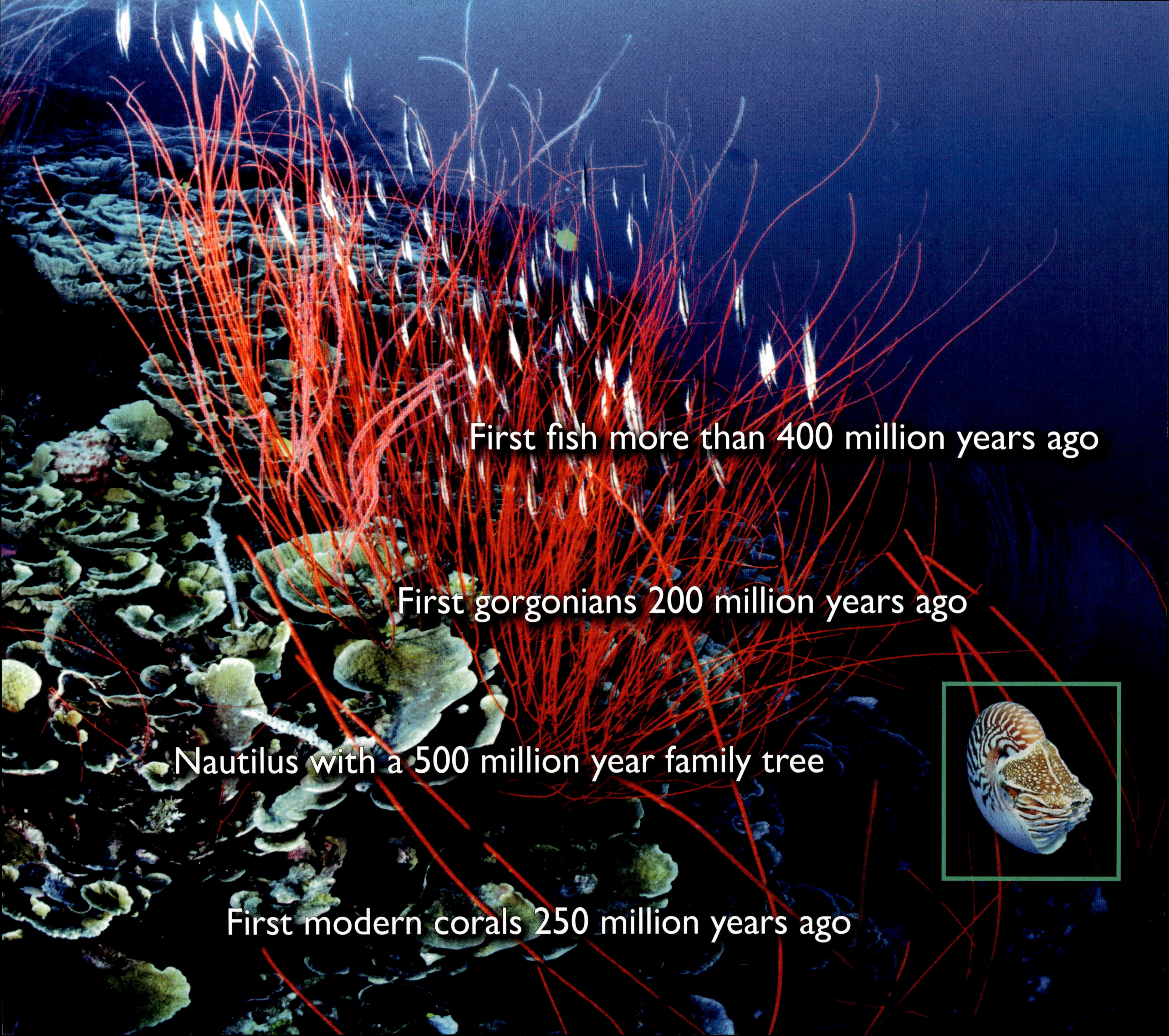
First fish more than 400 million years ago
First gorgonians 200 million years ago
Nautilus with a 500 million year family tree
First modern corals 250 million years ago

CONTINUITIES

Celebrating life in the eternal ocean

Scuba diving in rock.

Extracting fossils from their surrounding rocky matrix is mostly an arduous task that involves high-tech tools and a great deal of dust. What has kept my spirits up working on some of the more difficult pieces is the realisation that I am uncovering things that no human eyes have ever seen. Wearing a mask to protect me from flying debris, I imagine myself to be scuba diving in rock. I try to anticipate what organisms may emerge from the dusty matrix even before they do.

This is me in my element – working on preparing a giant Jurassic nautilus fossil from Dorset. Each step is a process of discovery, exposing traces of animal life 180 mya. It's not unlike scuba diving in rock.

Scuba diving on ancient reefs.

Actually, it is more than just imagination. If you snorkel across a tropical coral reef today you are ACTUALLY seeing a vista that would have been common in Jurassic times, and even earlier. Many corals, feather stars, sponges, gastropods and fish have changed so little that an ichthyosaur (or you) would be hard pressed to tell the difference – so is the great continuity of life on the seabed.

This almost incredible continuity of ocean life has not been given much prominence in the popular media, which chooses instead to focus more on possible extinctions than those that won't go that way. Mayhem and murder (and extinction) always get more media attention than peace and tranquillity (and continuity).

Extinction is often fashionably top of mind in the modern world, as is the human role in its apparently inevitable process. Sometimes we seem to relish living at the heart of entropy and seldom focus on those groups that survive and continue to thrive.

If you want to know how to survive a catastrophe – ask any Nautilus. They and their ancestors have thrived for 500 million years and lived through five mass extinctions.

Lineages may wax and wane – there were hundreds of slit shell species in the Jurassic, for example, but only about thirty exist now – but the complete termination of an evolutionary path and the demise of all its representatives is relatively rare compared with the number of things that have continued to this day.

Surely, survival is as interesting a phenomenon as extinction?

This sea urchin's body plan has been almost unchanged for 450 million years.

In this chapter we'll focus on that kind of continuity and show that the ocean is an environment that tends towards *continuity*, and not *change*. We'll look at the beauty of today's marine life and contrast it against its fossil ancestors – imagining all the time the bold colours that would surely have been behind these somewhat dull stone fossils.

Of the 36 recognised phyla of living animals, this chapter celebrates marine animals selected from the nine phyla highlighted below.

Phylum	Inventory of Animal Species	Number of named species (incl. synonyms)
Arthropoda	Insects, arachnids, crustaceans	1,214,000
Mollusca	Cephalopods, gastropods, bivalves etc	117,000
Chordata	Fish, birds, lizards, amphibians, mammals	68,000
Platyhelminthes	Flat worms	29,000
Cnidaria	Jellyfish, sea anemones, corals	10,000
Porifera	Sponges	8,700
Echinodermata	Sea stars, sea urchins, crinoids…etc	7,500
Bryozoa	Bryozoans	5,700
Brachiopoda	Brachiopods	440
	Other worms	42,000
	Other animals	4,000
	Total number of named animal species	1.5 million
	Of which about 30% are found in the ocean.	498,000

Source: MarineSpecies.org and CatalogueOfLife.org

There are 36 recognised phyla of living animals – all were born of the ocean and have representatives living there today. All have ancient histories.

The phyla covered represent the vast majority of common animals you are likely to encounter in a casual exploration of today's ocean. For reasons of space I have chosen to include only the flatworms and omit 42,000 other species of the worm-kind – they will have to wait for another opportunity.

Many living things around us have been on earth in recognisable form for hundreds of millions of years, creating a continuity of life that we sometimes fail to appreciate. This is their story.

Giant barrel sponges,
two meters tall,
in deep water off
Papua New Guinea.

Flower basket sponge *Euplectella aspergillum* illustration from the *Challenger* deep sea expeditions 1872–1876. Known as glass sponges the skeleton and roots are made of spicules of pure silicon.

A bright red sponge provides a spectacular perch for a poison goby *Gobiodon citrinus*. It is named for its poisonous mucus. Ras Muhammad, Red Sea, Egypt.

When put through a sieve to separate all the cells, sponges have the uncanny ability to re-assemble themselves.

Phylum Porifera
SPONGES

Sponges are some of the simplest multi-celled beings. They live rooted to the sea floor, don't move about and lack a brain, head, eyes, mouth and almost everything else you would associate with an *animal*.

They are also amongst the oldest animals, considered by some to have preceded all others. Sponges appeared before the Cambrian, perhaps even 750 million years ago, during the *Snowball Earth* period. They can live in waters with very low oxygen concentrations and may have been instrumental in oxygenating the ocean.

Sponges contain hard silica or calcite spicules that support their structure, and these are often the only parts that remain as fossils.

Sponges feed by filtering miniscule particles of food out of the water. They just sit there, almost inanimate. Their larval off-spring however became quite active, searching for new places to settle.

Some sponges grow to massive sizes and are estimated to live for far beyond 2,000 years. In the Caribbean and around Papua New Guinea I have dived amongst barrel sponges that are two metres tall with bases bigger than four oil drums. Encrusting sponges can cover huge areas of rocks, or metal surfaces of shipwrecks, with their beautiful hues – silent and immovable, still thriving.

Despite their simplicity, sponges have a level of cellular organisation that puts other life forms in the shade. If one separates their cells (for example by passing them through a sieve) and then places them back into sea water, over a few days they will re-assemble themselves into small sponges again. No other animal can resurrect itself this way.

Could sponges be the ancestors of all complex life?

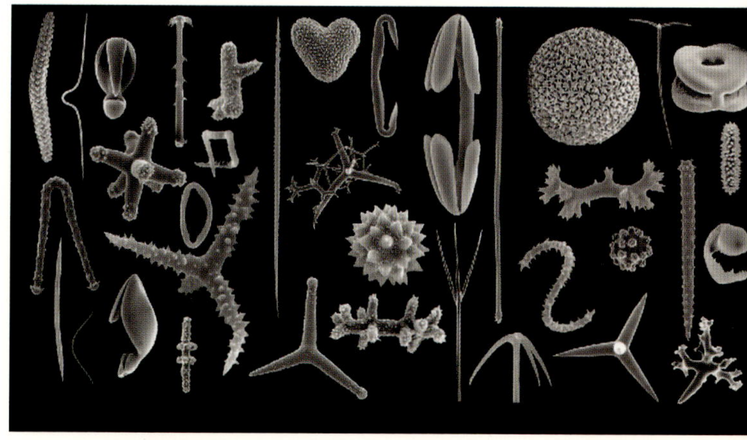

Scanning electron micrograph images of a selection of sponge spicules not to scale, sizes vary between 0.01 and 1 mm. These are the only parts that fossilise.

An orange encrusting sponge, possibly *Crambe crambe* or *Acarnus* sp. growing around a purple hydrocoral *Allopora* sp. Ras Muhammad, Red Sea, Egypt.

Phylum Cnidaria
JELLYFISH & CORALS

Brain coral spawning.

Look at any healthy coral reef and the phylum Cnidaria will be extremely well represented, even to the casual observer. Not only are they prolific today, but this phylum includes some of the most ancient marine invertebrates. They are broadly grouped into those with sessile polyps attached to hard substrates (Anthozoans, such as the corals) and the swimming medusae (Medusozoans, the jellyfish and hydrozoans).

Together Cnidaria includes more than 10,000 living species and a fossil record extending way back to before the Cambrian explosion of life.

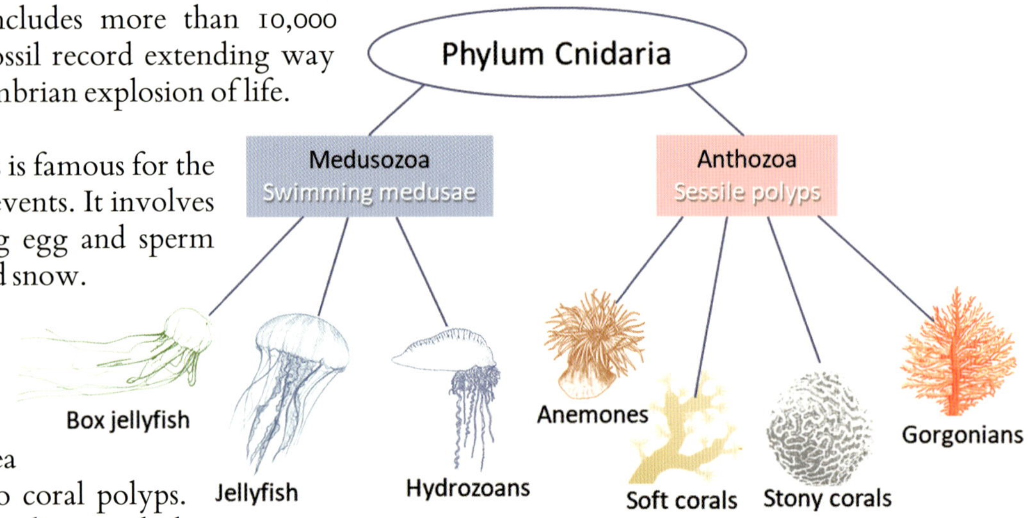

The reproductive cycle of adult corals is famous for the annual synchronised coral spawning events. It involves coral polyps simultaneously releasing egg and sperm bundles in a blizzard of multi-coloured snow. This synchrony is crucial, because the eggs are viable for fertilisation for only a few hours. After fertilisation, the larvae are distributed as part of the plankton and then settle on the sea bed, undergoing metamorphosis into coral polyps. Although closely related to jellyfish, anthozoans lack a medusa or *jellyfish* stage in their life cycle.

Sea anemones are mostly solitary, while most corals are colonial, being formed by the budding of new polyps from a founding individual. Colonies of stony (hard) corals are strengthened by calcium carbonate and other materials and take various plate-like, bushy or leafy forms that are often the primary structure of a coral reef. Many of the animals in this phylum create truly inspiring microworlds of commensal relationships, as we will see in the following pages.

Jellyfish

Jellyfish, also known as *medusae* (singular *medusa*), have roamed the ocean for at least 500 million years, and possibly 700 million years or more, making them the oldest known animals to have organized tissues and a nervous system. They're also the first marine animals known to have begun swimming using muscles instead of just drifting with the currents.

Jellyfish are likely the oldest multi-organ animals.

The popular name 'jellyfish' was reportedly coined in 1796, and the term has stuck to encompass any free-swimming soft-bodied animal with a jelly-like umbrella-shaped hood and trailing tentacles. They are found world-wide and at all depths in the ocean. Jellyfish are a classical example of the continuity of life.

The problem is that jellyfish are primarily *soft* animals and generally leave little fossil evidence that they ever existed. However, there are exceptions, such as the Cambrian fossils found in Utah on which you can see a distinct bell-shape, tentacles and muscle scars.

The hood typically pulsates for propulsion and the tentacles may be used for defence or to capture prey. Some species contain powerful toxins to assist in both these objectives and are often portrayed as "fearful denizens of the deep". This reputation is largely unfounded as most are quite benign.

Having said that, there are some notable exceptions, representatives of which are best avoided, such as the large box jellyfish. It is famed for its venom, amongst the deadliest in the world, containing powerful toxins that attack the heart, nervous system, and skin cells and can be fatal to humans.

Jellyfish are made of soft tissue and seldom leave any fossil evidence. This rare Cambrian fossil from Utah is believed to be 505 million years old.

Life-size replica of a Lion's Mane jellyfish and a thresher shark for size comparison in the Smithsonian Institution, Washington.

Today's giant amongst jellyfish is the Lion's Mane *Cyanea capillata*, the largest recorded specimen of which was found washed up on Massachusetts Bay in 1870 and had a hood with a diameter of 2.3 metres and tentacles 37 metres long. Certainly big enough to fuel fertile imaginations.

This jellyfish features in the Sherlock Holmes short story *The Adventure of the Lion's Mane*. A professor dies under mysterious circumstances shortly after going swimming, shouting "the lion's mane" before he succumbs. Suspicion was originally laid upon the professor's love rival, until the latter is similarly attacked, but survives, although badly stung. Holmes subsequently discovers that the true killer is this jellyfish. In the context of the story, it is only because the professor has a weak heart that he succumbs, as is confirmed by the survival of his rival.

In the real world though, even these giants are not fatal to humans and their irritating stings can be easily treated with vinegar.

Much worse is the Portuguese-Man-of-War *Physalia* sp., variously known as a *Blue Bottle* or *Floating Terror*. It is not a true jellyfish but a marine hydrozoan, a colonial organism made up of specialized individual animals. Its tentacles deliver a powerful sting that is feared by coastal bathers.

Encountering large and often dramatically colourful jellyfish on a dive is a memorable event as many play host to commensal and parasitic passengers such as crabs, shrimps, fish and brittle stars hitchhiking amongst the tentacles.

Close-up of a Lion's Mane jellyfish with a white brittle star passenger. Pemba, Mozambique.

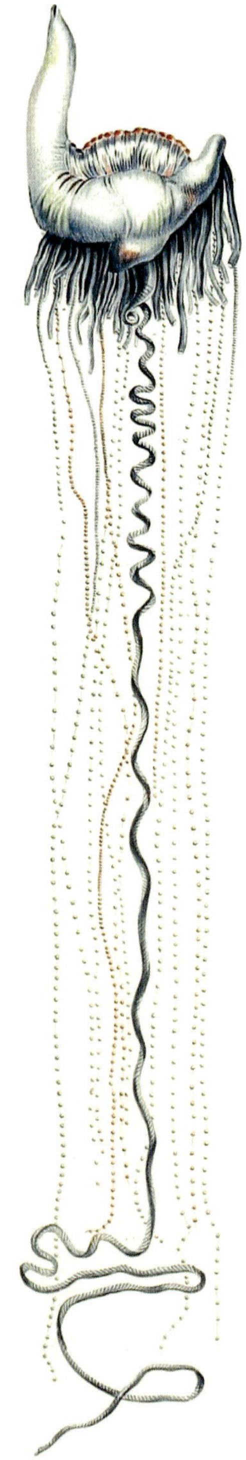

Illustration of *Physalia physalis*, in Lesueur *Voyage de découvertes aux terres Australes* 1807.

The upside-down Jellyfish *Cassiopeia frondosa*, as its name implies, usually lives upside-down on sandy bottoms, but I have also found it found swimming upside-down in mid-water, sometimes with passengers such as this crab.

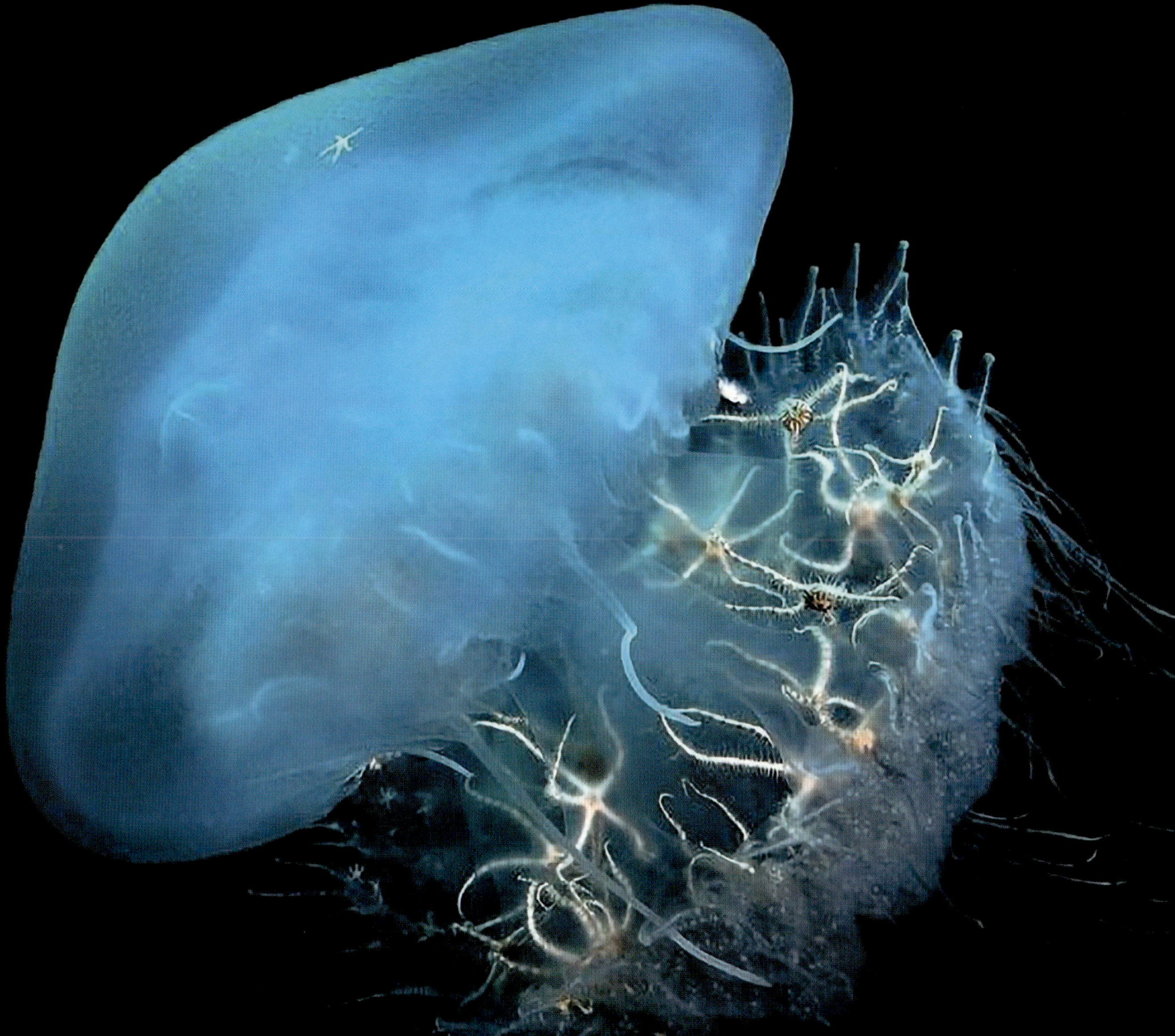

This moon jellyfish *Aurelia aurita*, bristling with an attendant over-population of brittle stars *Ophiocnemis marmorata*, was photographed off the east coast of Africa, near Mozambique, in the Bazaruto Archipelago. It is presumed that the brittle stars initially settle on the jellyfish as planktonic larvae and subsequently steal food from their host. It is interesting to see a tiny juvenile perched atop the jellyfish bell.

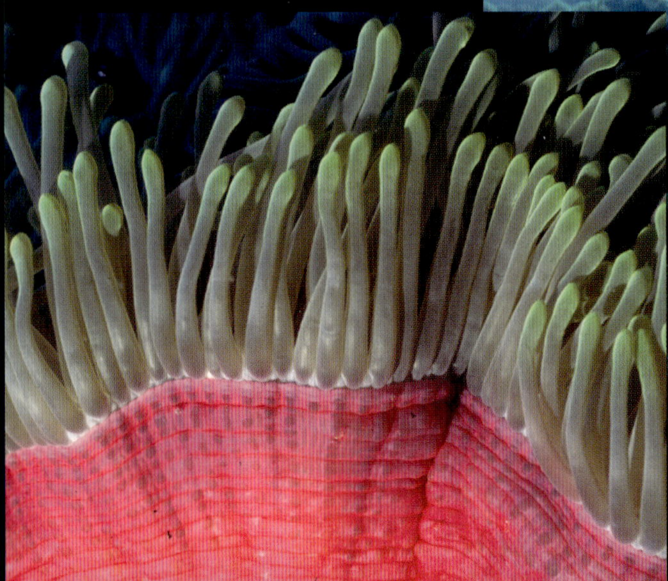

Sea anemones

Sea Anemones are named after the anemone, a terrestrial flowering plant, because of the superficial similarities in their appearance. They are typically made up of a single polyp attached to a solid surface, with a plethora of tentacles around a central mouth. They can retract their tentacles completely into their bodies when threatened and are carnivores, catching prey with their tentacles. Some species get supplemental nourishment from algae living in their cells.

Many species are armed with stinging cells. Despite this they typically live in close association with families of clown fish, crabs and shrimps, all of whom seem to be immune to their poison. The typical sea anemone is a veritable microworld of commensal animals with some species so selective that they limit their association to only one species of clownfish.

Because sea anemones have no hard parts there are few fossils. But an elongated bag-like animal *Mackenzia* found in the Burgess Shale is believed to be the oldest ancestor at more than 500 million years old.

Main picture: Sea anemones and soft corals crowd atop a rocky reef outcrop. Astove Island, Indian Ocean.

Smaller pictures show a range of commensal relationships with sea anemones – clown fish, crabs and shrimp – photographed in the Indian and Pacific Oceans.

This rare sea anemone *Korsaranthus natalensis* is unusually mobile – it moves in the manner of a *Slinky* toy and can sometimes be seen using its parachute-like walking disc to sail to new locations. It preys on octocorals and sports distinctive colours – red or orange with white stripes and is easily distinguished in the field. This rare anemone is so far known only from South Africa – this one photographed at a depth of 40 m on Landers Reef, Kwa Zulu Natal, South Africa.

Soft Corals

Soft corals are best-known for forming flexible tree-like structures, sometimes of giant size. During the day these corals look like insignificant deflated clumps on the reef and only show their brilliant colours when fully inflated with the tentacles exposed for feeding at night.

The polyps of soft corals have eight feathery tentacles. Grouped within the subclass Octocorallia (the octocorals), they number more than 3,000 living species and include sea pens and gorgonians.

Unlike their hard or stony coral relatives, octocorals do not produce calcium carbonate skeletons but contain minute, spiny skeletal elements called spicules or sclerites – they can be seen inside the white transparent stems in the photo top right. While providing support, spicules also give their flesh a spiky texture which is believed to deter predators.

Inflated soft coral tree resplendent at night, clearly showing eight-tentacle structure and spicules inside transparent arms. Kimbe Bay, Papua New Guinea.

Sea pens are named for their supposed resemblance to antique quill pens and show this form when extended into currents to feed.

Because they lack a hard skeleton octocorals are rarely found as fossils, but many palaeontologists believe that one of oldest known animals, *Charnia*, found in Precambrian rocks, belongs to this group. Some tiny needle-like Cambrian fossil spicules have been found that appear similar to the spicules of living octocorals, suggesting that the ancestry of these animals goes back more than 500 million years.

Fossil and reconstruction of *Charnia masoni* more than 600 million years old.

Sea pens extended into the current for feeding at night.

Gorgonians

Sea whips, a kind of gorgonian, on a sea mount in Papua New Guinea.

Gorgonians, sometimes called sea fans and sea whips, are found throughout the world's tropical and sub-tropical oceans. They are closely related to corals, having individual polyps in colonies on a flat, branching structure composed of intermeshed spicules of hard calcium carbonate, reminiscent of a hard fan. Others may be like tall thin whips. Gorgonians are the source of coral used for jewellery.

The valuable red coral or *Precious Coral* has been desired in the jewellery trade from the Middle Ages and was originally found mainly in the Mediterranean Sea. It is from the gorgonian *Corallium rubrum* (formerly *Gorgonia nobilis*) of which the supplies in shallow water have been totally depleted in that area. The industry has moved to the Pacific where the black variety is also found. There are about 500 different named species of gorgonians, of which about half are the source of black coral.

Sea fans and sea whips play host to a variety of marine life such as the cryptic hawkfish in the image on the facing page. The superb camouflage helps protect it from predators and prepare for the occasional snack.

Gorgonian tissues are the primary home of flamingo tongue gastropods *Cyphoma* sp. (family Ovulidae, sometimes called the false cowries) and they feed on its tissues and polyps. Gorgonian polyps are also a preferred diet for some species of butterflyfish.

Giant sea fans, each more than two meters across crowd on a reef wall while a huge school of fish sweeps upwards past it. Daram island, southern Raja Ampat, Indonesia.

Contrasting colours of a red gorgonian with white polyps provide excellent camouflage for a cryptic hawkfish *Oxycirrhites typus*. Bismarck Sea, Papua New Guinea.

A pair of *Cyphoma mcgintyi* feeding on a gorgonian.

Butterflyfish feeding on gorgonian polyps, Red Sea.

A giant humphead parrotfish *Bolbometopon muricatum*, more than one meter long, feeding on hard coral polyps and their calcium carbonate skeletons. Parrotfish are said to create much of the coral sand on tropical beaches. Praslin, Seychelles.

A staghorn coral tree *Acropora cervicornis* surrounded by Purple Queen anthias *Pseduanthias pascalus* on a healthy reef in Milne Bay, Papua New Guinea.

Stylaster sp., the so-called 'precious coral' growing on the roof of a cave. Mayotte, Comoros Islands, Indian Ocean. Size about 150 mm.

Hard (or Stony) Corals

Today, most hard corals are colonial. The individual animals, the polyps, have a cylindrical body with a mouth fringed by tentacles, usually in multiples of six, which are not feathery as in soft corals. The founding polyp secretes calcium carbonate in the form of aragonite crystals to protect its tiny soft body. The polyps reproduce by budding, forming a colony with a shared skeleton that may be up to several metres across. They are the dominant reef-building organisms in tropical marine habitats. Corals are carnivores, catching prey with their tentacles. Many species supplement their energy needs by making use of photosynthetic single-celled algae that live within their tissues.

The early rugose corals were not true reef-builders but were solitary individuals, although they often lived in very large groups. They and the tabulate corals were constructed of light and porous cups of calcite, rather than the aragonite of hard corals today. All these were wiped out in the mass extinction at the end of the Permian Period 251 million years ago.

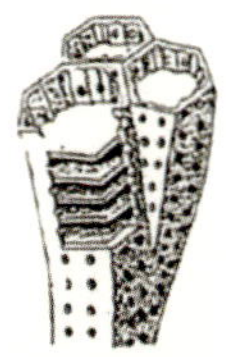

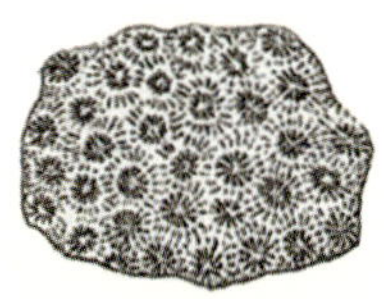

Hard coral	Rugose	Tabulate	Scleractinian
Period	Ordovician - Permian	Ordovician - Permian	Triassic - Recent
Symmetry	Bilateral	Radial	Radial
Community	Colonial & solitary	Colonial	Colonial & solitary
Skeleton	Calcite	Calcite	Aragonite

The ancestors of today's hard corals first appeared in the Middle Triassic 225 million years ago, about 25 million years after the Permian extinction, but their relationship to the corals of the previous eras is currently still uncertain. These scleractinian hard corals rapidly expanded into ecological niches once dominated by their predecessors. They were at their greatest diversity in the Jurassic and all but disappeared at the end of the Cretaceous, with only 18 out of 67 genera surviving. Since then they have again

Fossil of the large solitary horn coral *Paleophyllum* sp. from the Ordovician of China 450 mya, size 260 mm; and a drawing from Ernst Haeckel's *Kunstformen der Natur* 1904.

Scleractinian coral head showing shared skeletons deposited by coral polyps – retracted during this daytime photo.

The fully exposed polyps of a stony coral, completely covering the stony skeleton, photographed at night.

The fossil coral *Holocystis elegans* sitting on top of a large bivalve *Mulletia mulleti*, Perna Bed, Atherfield, Isle of Wight, UK. 120 mya.

Thecosmilia annularis. A fossil colonial coral made up of a number of polyps. 200 mm high. Steeple Ashton, Wiltshire, UK. Jurassic 155 mya. Surrounded by several examples of individual coral cups.

flourished to become the dominant reef-building organisms in tropical marine habitats, perhaps as a result of teaming up with symbiotic algae.

Corals are extremely sensitive to changes in light, temperature and salinity, and their survival is thus easily threatened by natural and man-made changes to the environment. Hard corals are the tropical ocean's equivalent of *canaries in the goldmine*. Stress on hard corals in the Indian Ocean from several El Niño events since 1997 killed off large populations and the damaged reefs were preyed on by exploding populations of black *Diadema antillarum* sea urchins feeding on the resulting blooms of slimy green algae.

Many reef animals live amongst the polyps of hard corals and some feed on the polyps themselves, notably a diverse range of butterflyfish.

A transparent shrimp *Urocaridella antonbruunii*, pregnant with eggs visible through its transparent skin, amongst the extended polyps of a solitary coral. Comoros Islands, Indian Ocean.

Solitary Corals

Solitary corals are different from their colonial counterparts in that a single polyp creates a solitary calcium carbonate cup. The skeleton is never attached to the substrate and retains its shape, the polyp never budding, but grows outwards often to large sizes. Some species may change sex, with small or stressed individuals typically being male. Sex change is a powerful evolutionary adaptation, allowing corals to survive during periods of stress.

The animal can retract completely into its stony skeleton and sometimes sandy ledges between reefs seem to be strewn with dead solitary coral skeletons, only to 'come to life' at night as the animal emerges. This is a spectacular sight, especially with the larger species whose single large polyp can be as large as 350 mm – such as this one in my collection – clearly showing why it has been called "the Godzilla of the coral world".

Solitary coral skeleton.

Skeleton with animal emerging.

Skeleton with animal fully extended – note this is a single polyp.

The stony cups of *Stephanocyathus spiniger* a solitary, deep-water stony coral that lives in 200–1,000 m depths. This one has six long spines that stop it sinking into the substrate. Size 30 mm.

A huge solitary coral skeleton 350 mm across and 180 mm high – found on a sandy bottom at around 30 m depth in the Indian Ocean near Mayotte. Unidentified species.

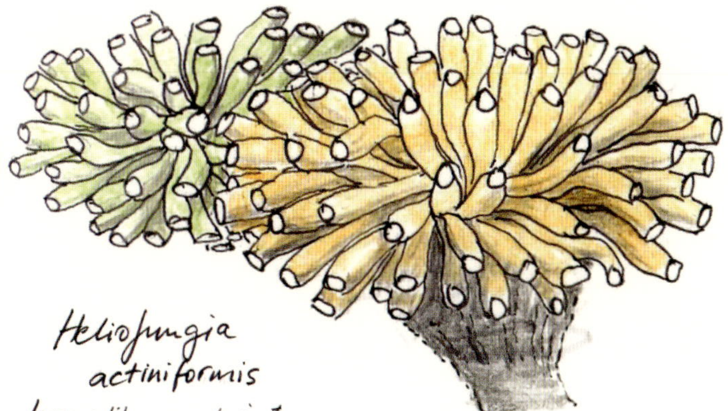

An example of the living pink encrusting bryozoan *Eurystomella bilabiata*:

Phylum Bryozoa

THE BRYOZOANS

This large group of small marine invertebrate animals represents filter feeders that sieve food particles from the water using a retractable crown of tentacles. The animals themselves are tiny, typically about half a millimetre long, and are mostly colonial animals that form fans, bushes and sheets. One group in this phylum, the Cheilostomata, produce mineralised calcium carbonate skeletons in single-layered sheets which encrust reef and other substrates. One of these, the so-called 'lace coral', has three-dimensional skeletons in the form of folded lace-like sheets. Fragments of these are often washed ashore after storms.

An example of an unusual bryozoan, the lace coral *Hippomenella vellecata*, fragments of which are often cast up on beaches. Size 50 mm.

Bryozoans are prolific today with more than 5,700 species. Their mineralised fossil skeletons first appeared in rocks from the Early Ordovician period but bryozoans without mineral skeletons may have lived much earlier without leaving any fossil record. They may have differed significantly from the fossil and modern forms we know.

The encrusting nature of bryozoans is perfectly illustrated by fossil examples such as this one (at left). Here a discarded shell of the nautilid *Aturia* has acted as a nucleus around which a thick bryozoan crust has developed. The whole has been fossilised together.

Fossil carbonate bryozoan concretion (called a bryolith) containing an *Aturia* sp. nautilid, found in Saint Martin d'Oney, France. Lower Miocene, 23 mya. Size of nodule 85 mm.

Bryozoans were the last major phylum to appear in the fossil record.

A fossil mineral skeleton of the bryozoan *Evactinopora radiata*. It lived free of attachment on the seafloor. Jefferson County, Missouri. Carboniferous Period 350 mya. Size 5 mm.

Phylum Echinodermata
THE ECHINODERMS

The phylum Echinodermata, a term that literally means *spiny-skinned animals*, consists of five well-defined clades containing some of the oldest complex animals, such as the crinoids, and some of today's most familiar marine organisms – from sea stars to sand dollars.

Crinoidea: Sea Lilies & Feather Stars
Ophiuroidea: Brittle Stars & Basket Stars
Asteroidea: Sea Stars & Cushion Stars
Echinoidea: Sea Urchins, Sand Dollars
Holothuroidea: Sea Cucumbers

There are around 7,500 named species thriving in the ocean today, plus an estimated 13,000 species in the fossil record. They have thrived for more than 500 million years and their five-sided symmetry remains an iconic feature in rock-pools and on coral reefs for beachcombers, snorkellers and divers. All echinoderms are marine, this is the largest phylum to lack any freshwater or land representatives.

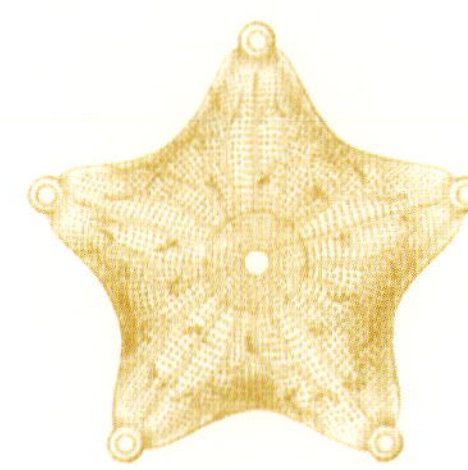

The bodies of echinoderms are primarily made of interlocking calcium plates, which, depending on the organism, can be tightly or loosely held together. Echinoderms have a system of internal water-filled canals, that connect to suckered tube feet, enable the animal to move, grip objects and feed. Echinoderms lack blood and achieve gas exchange via their water-based vascular system.

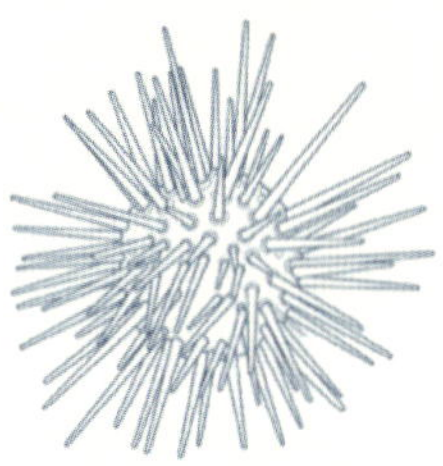

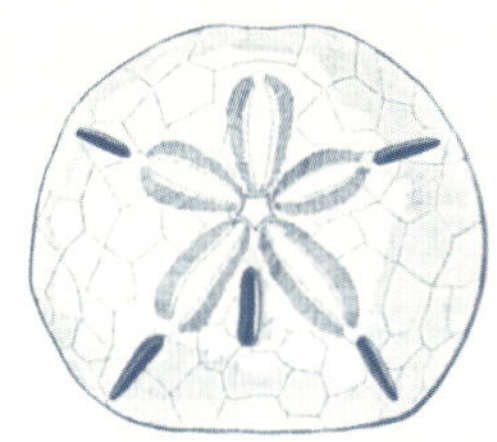

Echinoderms have thrived for more than 500 million years. Their five-sided symmetry remains an iconic feature in rock-pools and on coral reefs.

Many echinoderms graze on algae, using their hard jaws to gnaw the food and then excrete the remains from a pore on top. Any spines are typically attached by ball-and-socket joints, so they can move in any direction.

All the creatures grouped under the umbrella of 'echinoderms' may seem to be very different – but they do have a lot in common, and they all have a radial symmetry and tube feet.

The relationships between the animals in the Echinodermata were eloquently illustrated by David Attenborough in a 2010 edition of the BBC series *Life on Earth,* when he discussed echinoderm evolution:

"If the head of a crinoid  *is dropped on its face it looks much like a starfish* *. This, thinned down, then looks like a brittle star* *. And if it thickens, turns its arms back in on itself, and grows spines, you then have a sea urchin* *. One group of these became elongated, lay down on its side to feed, and became a sea cucumber* *".*

The business end of an unusual sea urchin showing a close-up of spines and tube feet, Jesser Point rock pools, Sodwana Bay, South Africa.

A rare sea urchin test found in situ on a reef. Sodwana Bay, South Africa.

Let's have a closer look at these intriguing animals, enjoy their idiosyncrasies and see how they relate to their fossil ancestors.

The evolution and changing lifestyles of sea lilies

The stalked crinoids, or sea lilies, look like plants but are actually complex animals with crowns of feather-like tentacles to catch floating food particles from atop flexible stalks. Their 'roots' can act as legs if the time comes to move. There are more than 600 living species and 6000 fossil species spanning a period over 500 million years.

All living crinoids are benthic filter feeders. When stalked *sea lilies* first radiated during the Ordovician Period between 485 and 443 mya, they developed diverse lifestyles to facilitate filter-feeding – even including pseudo-planktonic lifestyles, travelling in the open ocean attached to floating logs or even growing their own floaters – literally creating their own *island lifestyles*. Fossils illustrating these changing lifestyles are included on these pages.

Graphic from *Constructional Morphology of Pelagic Crinoids* Seilacher & Hauff 2004.

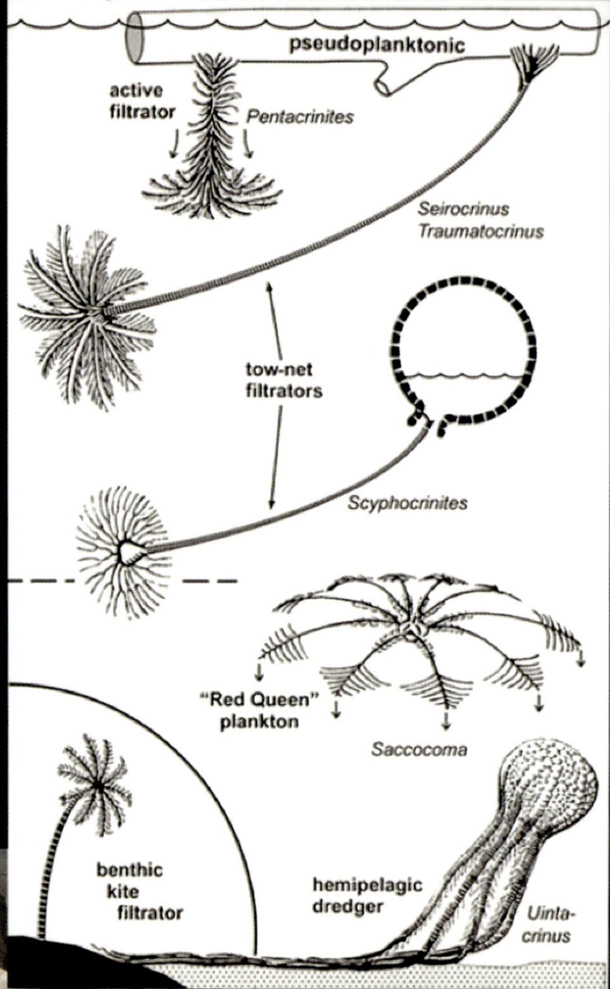

Late Silurian-Early Devonian: Fossil planktonic crinoid *Scyphocrinites*, and its loboliths, the multi-chambered floaters. Insets: Base and cross-section of a fossil lobolith showing crystals formed inside the flotation chambers during fossilisation. Locality: Erfoud area, Morocco. 420–410 mya.

Triassic: A large fossil community of pseudo-planktonic crinoids *Traumotocrinus xingyi* attached to a floating log. Locality: Guizhou Province, China. 230 mya.

Jurassic: This fossil crinoid community is made up of more than twenty specimens with complete heads and stems, all attached to driftwood. There is still some debate about the taxonomic relationships between various planktonic crinoids and it has been suggested (*Fossilien* Nov/Dec 2007) that this occurrence represents an intermediate morphology – some immature specimens resemble *Pentacrinites fossilis*, while some mature specimens resemble *Seirocrinus* sp. Locality: Watchet, Somerset, UK. 190 mya.

Cretaceous: A community of fossil stemless crinoids *Uintacrinus socialis*. They were originally thought to be 'free-floating' in the open ocean with their ten long arms, most more than one meter long, hanging downwards to feed. Now, however, they are considered by some to have been benthic animals inhabiting soft muddy bottoms, with the head buried in the mud and the long arms extended upwards to feed. According to Hans Hess, these colonies may have resembled "dense patches of eel grass". Locality: Niobrara Formation 83 mya, Smoky Hill Chalk, Gove County, Kansas, USA.

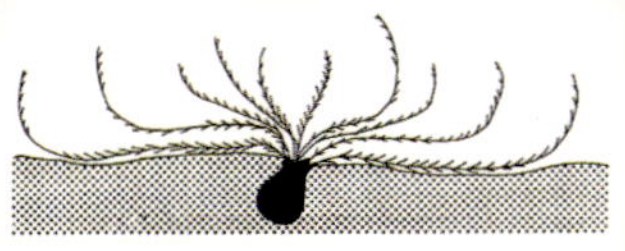

Milson et al 1994

Jurassic: A large community of pseudo-planktonic crinoids *Seirocrinus* sp. The stems are up to several meters long and are attached to a single 12 m log at the Urwelt-Museum Hauff in Germany. Locality: *Posidonienschiefer*, Holzmaden, Germany. 185 mya.

A fossilised community of stalked benthic crinoids *Encrinus liliiformis*. Muschelkalk, Alverdissen, Germany. Size 1.56 m, Triassic 230 mya.

Crinoids: Sea Lilies & Feather Stars

Those crinoids which are attached to the sea bottom by a stalk are commonly called stalked crinoids or *sea lilies*. The unstalked forms are called *feather stars*.

Relationships between the living forms of sea lilies and feather stars are still being debated. All living species are currently placed in Articulata, which is generally accepted as the only crinoid group to have survived the Permian–Triassic extinction event. New DNA evidence places some stalked crinoids within the feather stars, however, so clearly whether they are stalked or not may not be the best differentiator, and revision is required.

Some of the stalked crinoids survive today, but now their habitat is limited to the deep ocean and, only occasionally, more accessible places, such as on the wreck of the *Kirk Pride* in 250 m of water just off Grand Cayman in the central Caribbean.

The *Kirk Pride* sank in George Town Harbour in 1976 and lay unnoticed until 1985, when it was discovered trapped between two pinnacles on the Cayman Wall. The ship subsequently dislodged and fell to the deep ocean floor. I was fortunate to see and photograph these stalked crinoids (Order Isocrinida) on the wreck from a two-man research submersible shortly before that happened.

A time traveller from 400 million years ago would easily recognise these animals.

A red stalked sea lily crinoid *Proisocrinus ruberrimus*. Depth of ~1 km, near Wake Island in the Pacific.

Stalked crinoids growing on the wreck of the *Kirk Pride* off the Cayman Islands.

The feather stars – the *unstalked* crinoids – represent most of the species that exist today. Their modified feet allow them to clasp firmly onto the substrate. Their usual rooted posture belies the fact they can be free-swimming. When they release their grip, they swim with a graceful undulating motion. Both crinoids and feather stars have also been seen crawling along the sea floor using their tentacles.

Feather stars are familiar to anyone who has ever snorkelled or dived on a coral reef and they always deserve a closer look as they harbour commensal animals including crabs, sea spiders, shrimp and fish. Some of these have evolved to mimic their hosts with perfect camouflage such as this clingfish living within the crinoid's arms and feeding on detritus.

Feather star partially curled up on a gorgonian near Farakulufushi in the Maldives (above) and (below) a feather star with clingfish camouflaged amongst its tentacles. Kimbe Bay, Papua New Guines.

Colourful feather stars on reefs in, Indonesia (top right) and Farakulufushi in the Maldives (below).

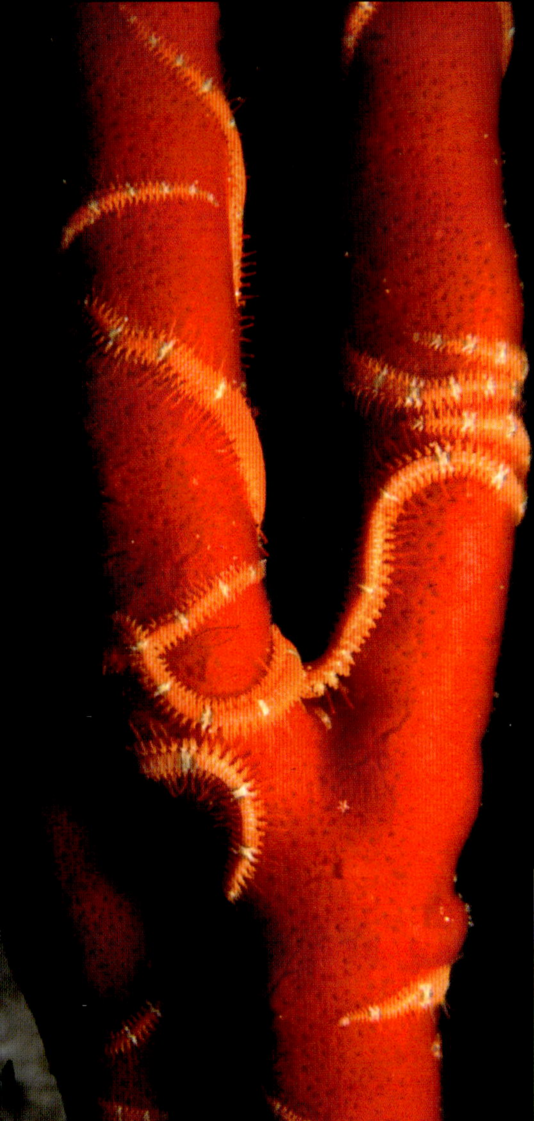

Today:

Brittle stars *Ophiothrix* sp. on a sponge at Sodwana Bay, South Africa, and on soft coral near Misool, southern Raja Ampat, Papua Province, Indonesia.

185 million years ago:

Fossil brittle star in matrix. *Palaeocoma egertoni*, Eype, Dorset, UK.

Brittle Stars & Basket Stars

Brittle stars and basket stars together make up the Class Ophiuroidea. They are primarily nocturnal feeders that hide during the daytime. More than 2,000 species are recognised today, with more than half found in deep waters below 200 m.

Brittle stars emerged almost 500 million years ago and have remained virtually unchanged since then. The fossil brittle stars in the conglomeration below are almost indistinguishable from those specimens that you might encounter in rock pools or on coral reefs today.

Basket stars are characterised by their improbable-looking fractal arms that can spread to a diameter of up to half a metre when feeding. They are covered in tiny hooks and spines which can be used to grip and manipulate food particles and move them down towards the central mouth.

The fossil record of basket stars is poor, but they are thought to have evolved from brittle stars during the Carboniferous Period, more than 300 million years ago.

Basket stars in life (top) and illustration (bottom image).

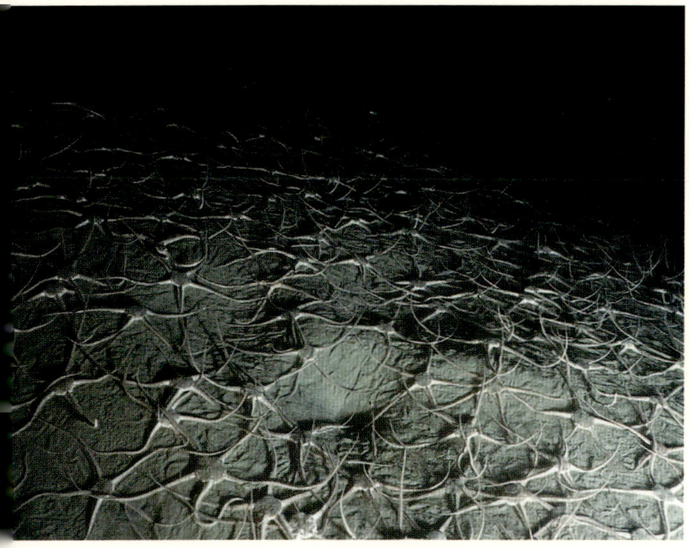

Brittle stars cover the seafloor at a depth of 600 meters Off the North Carolina coast, USA.

Fossil brittle stars *Palaeocoma egertoni*, together with small ammonites, a gastropod and a partial crinoid head. Eype, Dorset, UK. Jurassic 185 mya.

150 million years ago
Fossil starfish *Pentasteria longispina*
plus two brittle stars
Jurassic. Weissenstein, Switzerland.

Sea Stars & Cushion Stars

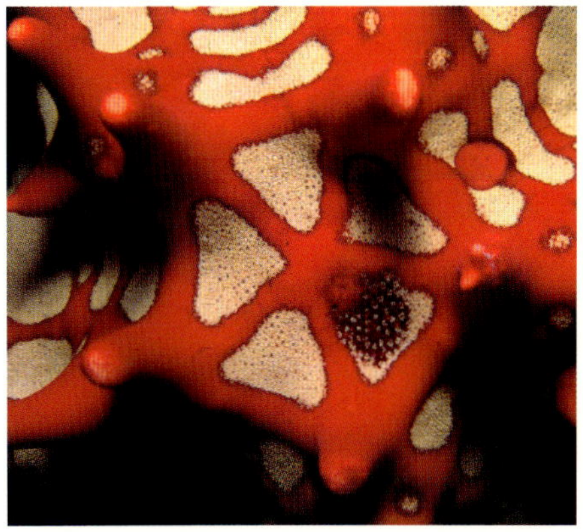

The top of a *Protoreaster linckii*, the African red knob sea star, showing the anus as a dark patch at bottom right on the central disc. Sodwana Bay, South Africa.

Sharing the five-pointed symmetry with the crinoids are their close relatives the sea stars and cushion stars.

The mouths of sea stars and cushion stars are underneath, and their anus is on top. They move on thousands of tube feet, each of which has a suction pad on the end, and which together are operated hydraulically via a network of water-filled vessels that run throughout the body just as in the sea urchins.

The cushion star gets its common name from its inflated, pillow-like appearance. Juveniles look like typical sea stars, but as they grow, they become more inflated and the arms grow together.

Both groups have been around since the Ordovician 450 mya.

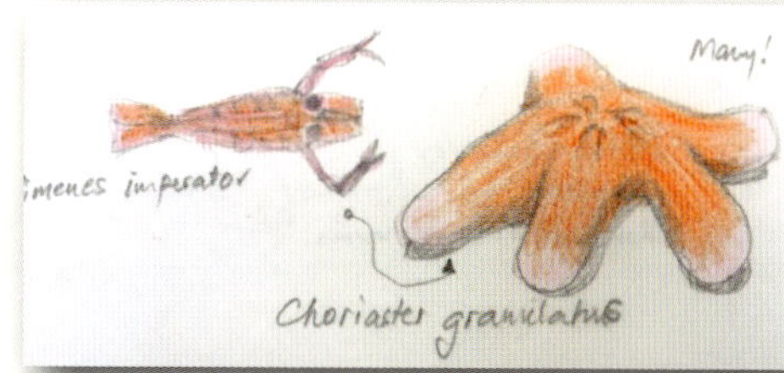

Today

Starfish *Astropecten articulatus*

A tiny 'mosaic' cushion star in a rock pool at St. Francis Bay, South Africa. Size 40 mm.

Sea Cucumbers

Sea cucumber. Perhaps a *Stichopus* sp. Comoros Islands, Indian Ocean.

The common name of the holothurians, the *sea cucumbers* refers to their resemblance to the popular salad ingredient. They are a diverse group of worm-like soft-bodied leathery echinoderms and are found on the sea floor worldwide. They use their tube feet to work their way along the sea floor, feeding on detritus, not unlike sea stars.

The large pineapple sea cucumber *Thelenota ananas* about 800 mm long, foraging under an overhang off Praslin, Seychelles, Indian Ocean.

There are about 1,700 species world-wide with the greatest number being in the Pacific. They are considered a delicacy in the Far East, which makes them susceptible to exploitation, even in protected areas such as the Galapagos Islands. Some diminutive species of sea cucumber may not exceed 10 mm fully grown and we have seen perpendicular reef walls in Astove Island in the Indian Ocean covered with hundreds of this yellow species. Some large species can reach a length of five meters.

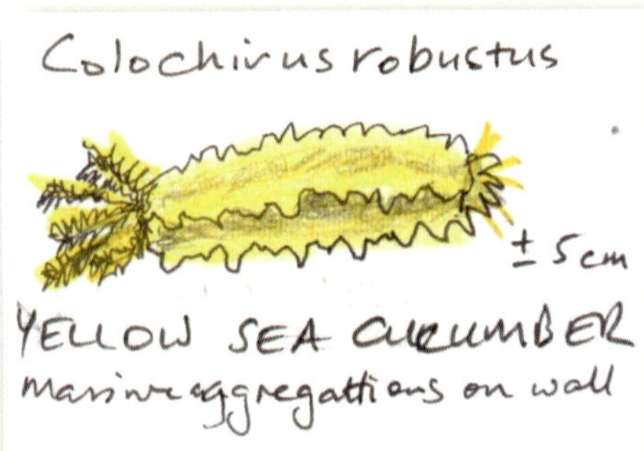

They are often associated with commensal animals – shrimp live on their skin and pearlfish use the rectum as a sanctuary from predators.

An astonishing feature of holothurians is their body wall, which is made of collagen – an abundant protein in human bodies that forms our bones, ligaments and skin. Sea cucumbers have evolved collagen with a unique property: they can rapidly 'loosen' or 'tighten' the tissues at will. If the animal wants to squeeze through a small gap, it can 'liquefy' its body and 'pour' through the space available. They can literally 'turn to jelly' temporarily, and then make the body firm again. The process was not understood until 2016 and is now being studied with the aim of developing treatments against skin ageing and for the manufacture of soft skins for human-like robots.

Shrimp on sea cucumber, Papua New Guinea.

Like all echinoderms, sea cucumbers have an endoskeleton made of calcified structures called ossicles (or sclerites) joined by connective tissue. It is through these that the oldest undoubted fossils of sea cucumbers have been identified, from the Silurian Period around 440 mya.

Fossilised calcareous inter-radial rings of holothurians. Size of largest piece 1 mm. Jurassic 180 mya, Germany.

Sea urchin crawling along sandy bottom with anal sac inflated. Milne Bay, Papua New Guinea.

Sea Urchins & Sand Dollars

Sea urchins are unmistakeable spiny inhabitants of the seabed, in all environments from the intertidal to the deepest ocean. They have a five-sided radial symmetry and many bear distinctive long spines that radiate from the body, or *test*. The variety of spines in different species is vast, both in living and fossil species.

They all move slowly, crawling on their tube feet, and sometimes pushing themselves along with their spines. Like their sea star cousins, their mouths are conveniently situated at the bottom to facilitate grazing and the anus is situated on the top. They feed primarily on algae but also eat slow-moving animals. They are predated on by sea otters, sea stars, sea otters, wolf eels and triggerfish.

About 950 species of sea urchin are known. *Sand dollars* are extremely flattened burrowing urchins, known as *sea cookies* in New Zealand and *pansy shells* in South Africa.

The hard shells and spines of sea urchins are frequently washed up on beaches after a storm and are counted as many a child's first 'sea treasures'. The same is true of fossil sea urchins – disarticulated parts are common but only in rare cases are complete specimens preserved. They have a rich fossil record dating back to the Ordovician, some 450 million years ago, and it takes expert knowledge to tell the shape of a fossil sea urchin spine from one living in the ocean today.

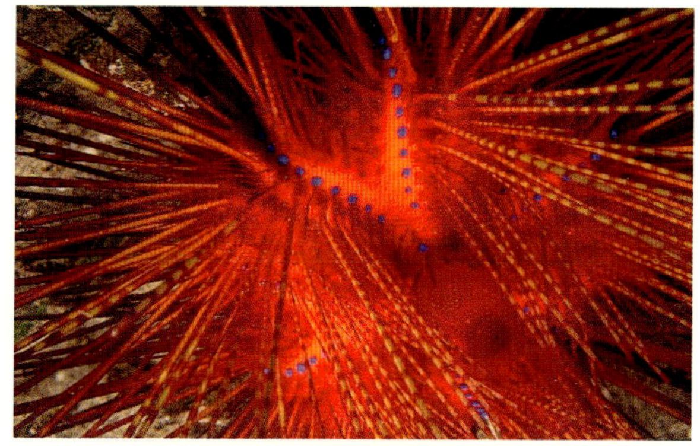

Blue-spotted Fire Urchin *Astropyga radiata*, Mayotte, Comoros Islands, Indian Ocean.

A rare unidentified sea urchin photographed at night on a reef at Pemba, Mozambique.

A selection of fossil spines of Cidaridae sea urchins.

This sea urchin taken off present-day St. Petersburg, Florida is almost indistinguishable from its counterpart in the Eocene of Paris, roughly 45 mya (on the right).

Phylum Brachiopoda

BRACHIOPODS

Almost 200 million years separate these two brachiopods. The earliest fossil brachiopods are found back in the Ordovician 450 mya.

The brachiopods are a group of animals with hard shells that mostly live the same stationary, filter-feeding life as bivalve molluscs, the clams, which they quite strongly resemble in shell shape and sculpture. The two groups are utterly unrelated, however. Most modern brachiopods attach to hard surfaces by means of a fleshy stalk called the *pedicle*. Also unique to the brachiopods is the *lophophore*, an organ utilised in both feeding and respiration. They are often known as *lamp shells* as their shells look somewhat like the pottery oil-lamps of the ancient world.

Their exact taxonomic classification is still the subject of much debate, but brachiopods have certainly been a unique group right back to the Cambrian. They were abundant for hundreds of millions of years but have gradually been displaced by the more successful clam shells.

Over 12,000 fossil species are recognized, grouped into over 5,000 genera. They are found today mostly in colder, deeper ocean environments but are thriving. There are around 100 genera in quite distinct families, and new species continue to be discovered. They are small animals – the largest living species is the size of a baby's hand, and some fossil species are huge, up to a metre or more.

Fossil brachiopod, 166 million years old, acid-etched to reveal interior structure including the lophophore used for filter-feeding. *Spiriferina pinguii* Jurassic, France.

Living brachiopod with interior structure clearly visible.

Phylum Platyhelminthes
THE FLATWORMS

Flatworms are regarded as a primitive stage in the evolution of animals with distinct front and rear ends.

This is a phylum of bilaterian, unsegmented, soft-bodied invertebrates. They are simple animals without a body cavity and without specialised circulatory or respiratory organs. Their flattened shape allows oxygen and nutrients to pass through their bodies by diffusion. Although simple they are extremely diverse, comprising more than 29,000 living species.

Free-living flatworms are mostly predators and are common in the ocean (especially in reef environments) and on land (in shaded, humid environments, such as leaf litter). Some (such as tapeworms) have complex life-cycles, with mature stages that live as parasites in the digestive systems of fish or land vertebrates.

In the ocean flatworms exhibit some of the most fascinating colours and patterns and are a favourite with photographers, along with nudibranchs (see further in this chapter under molluscs) with whom they are often confused. Flatworms have no gills – although some do appear to have flaps that look like ears, not unlike nudibranchs. Flatworms are sometimes seen swimming with a graceful undulating motion.

Two views of Bedford's flatworm *Pseudobiceros bedfordi*, one crawling along the seafloor and another swimming through a cloud of detritus. Raja Ampat, Papua Province, Indonesia.

Two specimens of the spectacularly patterned flatworm *Pseudobiceros bedfordi*.

Phylum Mollusca
MOLLUSCS

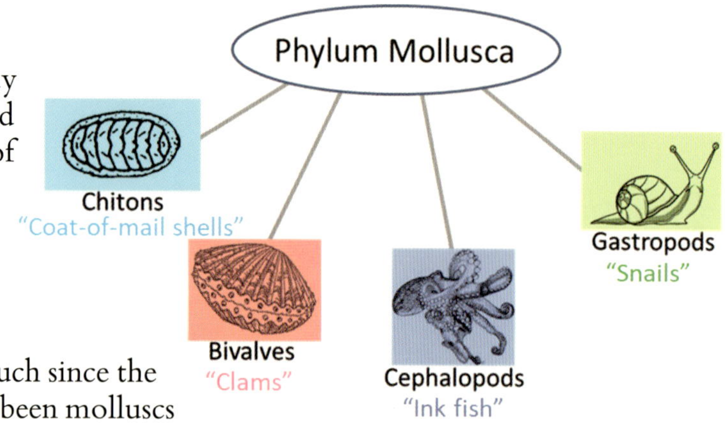

Mollusca is a large phylum of invertebrate animals that today comprise a quarter of all named marine organisms with around 117,000 living species named so far and a similar number of fossil ones. Of these marine invertebrates, only two major groups, the ammonites and the belemnites, are extinct. With a first appearance in the Cambrian Period more than 500 mya, molluscs epitomise the continuity of life.

Molluscs are the epitome of the continuity of life in the ocean, and the enormous variety that evolution tends to create.

For the last five hundred million years – pretty much since the advent of life forms that could move – there have been molluscs in the ocean. Gastropods (the snails) and bivalves (the clams) are the two largest groups. Both have since spread from the ocean into fresh water, and the snails alone onto land. Many snail species inhabit freshwater and terrestrial environments – and include, of course, the common garden snail.

Cephalopods too were amongst the first complex life forms in the Ordovician Period and are still prolific and thriving in the oceans. They today include some of the most intelligent animals around (the octopus and cuttlefish), the largest invertebrate on the planet (the giant squid) and perhaps the *least evolved* animal ever (the Nautilus) with a shell shape that has remained essentially unchanged for almost 500 million years.

Chitons may be a lesser-known group amongst the molluscs, but these so-called *coat-of-mail shells* are common in rock pools and shallow marine environments world-wide, with some inhabiting deep water up to 6,000 meters. Their shells are composed of eight separate overlapping plates that provide protection and permit the chiton to flex for locomotion, and even curl up into a ball. Some have fascinating characteristics such as light- sensitive shells (with thousands of tiny pores, at the base of each of which is a cluster of photosensor cells) and steel-tipped teeth.

An example of the complex shell forms of the genus *Ecphora* from the Miocene of Maryland. This entire genus became extinct more than five million years ago.

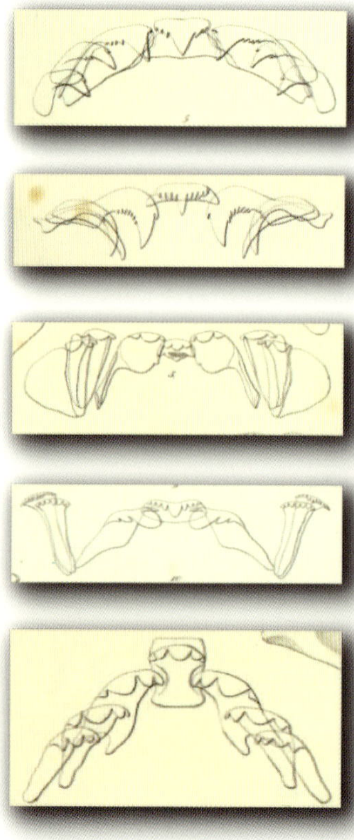

Selected illustrations of radulas from Troschel's *Der Gebiss der Schnecken* 1856–63.

A portion of a radula showing central teeth. Size of image is 40 µm.

The secrets of our success

Two of the secrets to the molluscs' success are the shell and the radula. Shells offer superb protection and are constantly built by an outer fleshy layer called the mantle which secretes the shell-building substances. The radula is a revolutionary feeding device that is common to the whole phylum except for the clams, including the extinct ammonites. In its most common form it resembles a chainsaw. Rows of sharp teeth are arranged on a flexible belt that the animal pulls back and forth over a fixed knob while pressing the whole thing against the surface of its food. Depending on the species, this can be an algae-encrusted rock face, a leaf or a dead fish. Known since the Cambrian, however, the basic radula has also been adapted for more specialized tasks, and particularly to facilitate the eating of meat.

The Organ Pipe coral (*Tubipora musica* at right) is an octocoral that forms a hard skeleton linking thousands of individual filter-feeding polyps in a dome-shaped colony. This lifestyle is shared by the colonial worm snail (*Petaloconchus innume* at left), despite the two being not even remotely related.

The shell of a fossil nautilus *Cymatoceras sakalavum* from Madagascar, 100 million years old, and a modern-day *Nautilus macromphalus* from New Caledonia. Size 180 and 162 mm.

The constant snails

What makes the snails so interesting is that they have constantly moved into new habitats without leaving the old ones, thus growing continuously in abundance and variety. They originated in the ocean and are still found there at all depths and in every niche. They moved into freshwater environments and are highly successful in estuaries, rivers, streams and ponds, and are now found everywhere on land too – from rain forests to alpine meadows and dry deserts. Snails make up about 85% of all living molluscs, which in terms of the number of species are second only to the insects.

Over their long history, the snails have adopted a dazzling variety of life styles. At the simplest level they have reverted to forming large colonies that look like corals and feeding by trapping minute particles of detritus in the water, just as many bivalves do. This stationary filter-feeding life makes the simpler molluscs both durable and abundant, but yet very vulnerable to large-scale changes in water quality.

At the other end of the scale are the predator snails – including the Cone Snails and their relatives, which prey on fish, worms and

The rare cone snail *Cylinder paulucciae* with its proboscis extended. Mayotte, Comoros Islands, western Indian Ocean.

Three male-female pairs of eulimid snails *Echineulima robusta* on a sea urchin *Diadema setosum*. The snails penetrate the urchin's hard shell and feed suctorially on its body fluids. Below is a cross-section of a starfish arm with a eulimid snail living inside it.

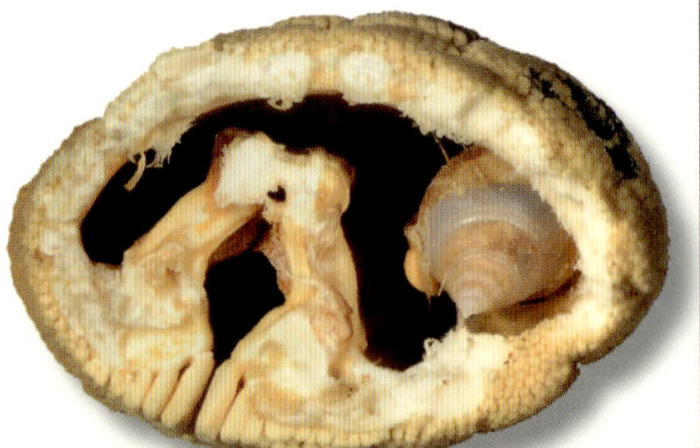

other molluscs. They have modified the radula into a highly sophisticated barbed hypodermic needle that they use to inject a cocktail of neurotoxins into their prey, rapidly paralyzing it.

Other snail groups have become parasites on their similarly ancient neighbours such as sea urchins and starfish, using a combination of the radula and acid to bore a hole through their host's shell or skin. In several groups the radula is used to drill through the shells of other molluscs, with the intention of then eating them alive.

We know perhaps more about the ancient history of the marine molluscs than that of any other animal group because their shells fossilise so frequently and so well. They often lie in fine marine sediment that quickly closes around them and facilitates fossilisation. The shapes and features of the different families found today remain recognizable far back in time, allowing us to track their evolution and pinpoint the relatively rare extinctions of whole groups.

Despite their huge age, two-thirds of all the snail families that have evolved so far, remain in existence today.

The living cockscomb oyster *Lopha cristagalli* from Madgascar, is barely distinguishable from its 170 million year old fossil ancestor *Actinostreon marshi* from the Jurassic rocks around Oborne, Dorset, UK.

A living pleurotomariid animal *Perotrochus quoyanus* from Roatan, Honduras. Collected at 215 m depth and photographed at 30 m amongst sponges.

The ancient Pleurotomaria

The family Pleurotomariidae is characterized by a notable slit on the outer lip of the shell which is used for expelling water. For obvious reasons they are commonly known as *slit shells*. They feed mostly on sponges though in captivity will eat almost anything.

This family was known only from fossil species until the 19th century when the first living specimen *Perotrochus quoyanus* was dredged in deep water off the West Indies by the "Blake" expedition of William Healey Dall in 1879. They were once considered rare but recent evidence suggests they are the dominant large invertebrates in their habitats.

They are the gastropod equivalent of the coelacanth – considered extinct but then found to be very much alive. This family is the oldest gastropod lineage with a fossil history going back more than 500 million years.

We know from the fossil record that they occupied shallow marine habitats world-wide until the end of the Cretaceous, after which they have been found mostly in deep water environments. Modern species are limited to deep-water habitats. Perhaps they, like the Nautilus, evolved into deeper water to escape the increasing diversification of predators such as fish and crustaceans and/or the acidification of the shallow waters after the extinction that killed off the dinosaurs and ammonites.

Throughout time there have been a great variety of shell forms in this family but the number was reduced in the Permian extinction. More than 1,500 fossil species have been described. Below is a selection of Jurassic fossils from the Bajocian about 170 mya demonstrating some of the diversity of this family at that time. The genus *Pleurotomaria* is now reserved exclusively for fossil species.

Below, left to right: *Pleurotomaria ornata* var *sulcifera* and *Pyrgotrochus bessina* both from Evrecy, France; *Pleurotomaria armata* from Bridport, Dorset, UK; *Pleurotomaria oxytera* from Sherborne, Dorset, UK; and *Obornella granulata*, from Oborne, Dorset, UK – this one is literally from our back garden.

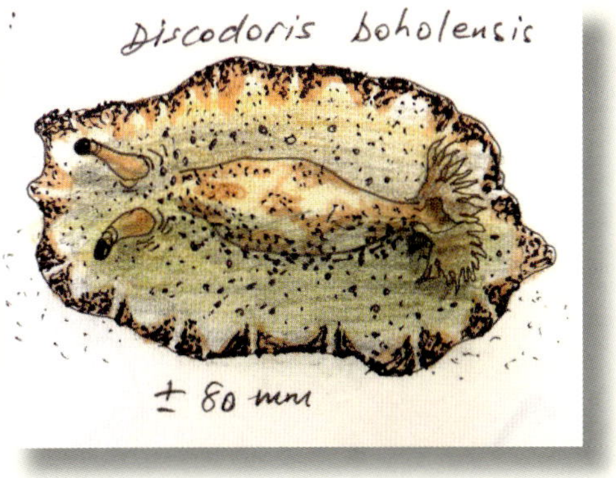

Nudibranchs: Naked snails

Nudibranchs (literally *naked gills*, after the exposed plumes of gills on the back of many species) are soft-bodied, marine gastropods which, as adults, lack shells. Their shelled ancestry is evident in their larval stage, when they bear a tiny shell, which is shed during metamorphosis.

One would think that jettisoning this most perfect protection would expose nudibranchs to many new dangers. They have however evolved new forms of defence and thrive today as one of the ocean's most successful and diverse invertebrate animals. They live in all seas, from shallow reefs down to depths of 700 m. More than 3,000 living species have been recorded so far, but this number is added to continuously.

Nudibranchs shed their hard shells during their larval stage to become naked snails.

They are well known for the extraordinary colours they display and as such are sought out by divers and photographers. The colours are believed to warn off predators against their stinging cells, which some have taken from anemones to use for their own defence.

Nudibranchs and land slugs do have a common ancestor, but a very distant one. The slug shape, literally shell-less snails, has evolved independently many times from shelled gastropods.

The giant Spanish Dancer nudibranch (so-called because of its undulating swimming motion in open water) *Hexabranchus sanguineus* is the largest known species in the world. It grows up to more than 400 mm in length but is seldom seen. It tends to emerge only at night to move somewhat conspicuously about the reef, making it one of my absolutely favourite night dive experiences – as shown in the photos on the facing page. This nudibranch lays thousands of eggs in a gelatinous ribbon which can appear like a magnificent rose on the reef. It is common for this species to have several commensal shrimps living amongst the folds of its mantle – all them so perfectly coloured as to be almost invisible. Can you spot one opposite?

Shown on the facing page are several varieties of the giant Spanish Dancer nudibranch and (above) its egg mass. Photographed at Sodwana Bay, South Africa; Pemba, Mozambique and on Aldabra Atoll in the mid-Indian Ocean.

A plethora of nudibranchs photographed in the Red Sea, Papua Province Indonesia and South Africa.

Cephalopods

All the familiar cephalopods in today's ocean have their origins in externally-shelled molluscs with a history that goes all the way back to the Cambrian more than 500 million years ago. Only the ammonites and belemnites are extinct.

Of the living cephalopods, only *Nautilus* has retained an external shell, while *Spirula* has evolved an internal shell that is almost identical to that of Nautilus, and reminiscent of the extinct ammonites.

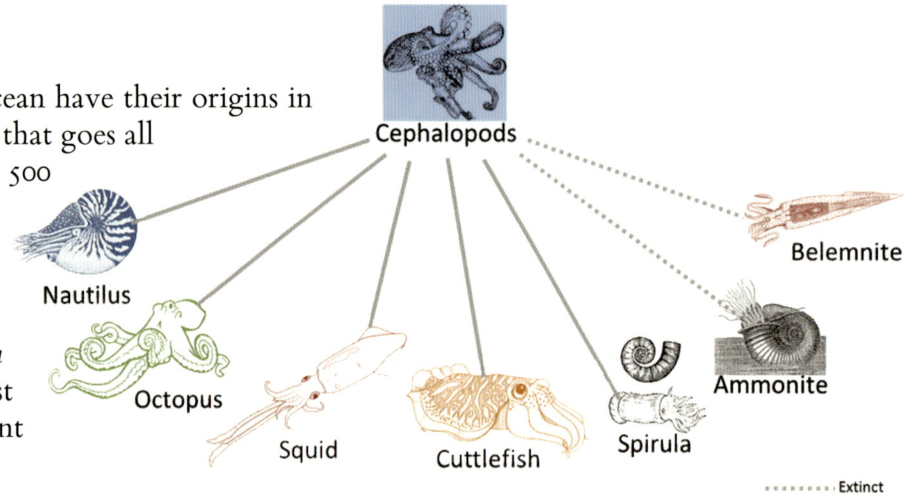

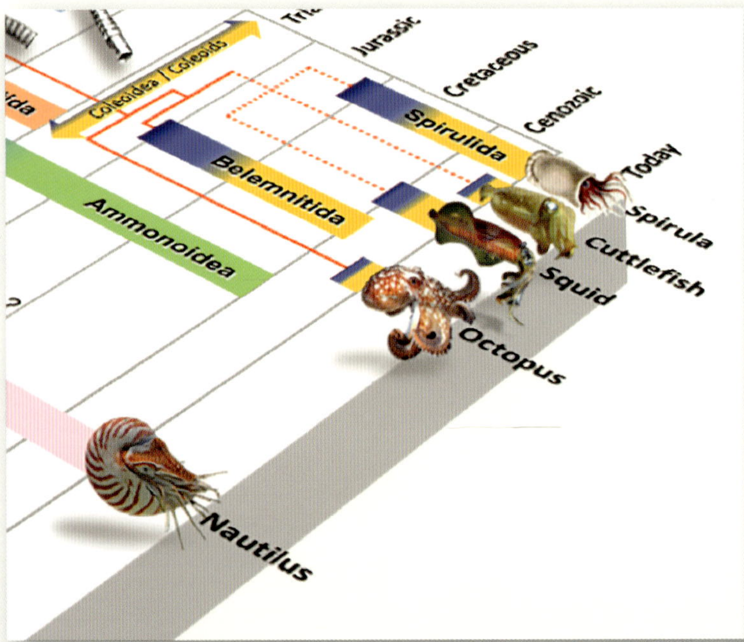

The 'soft' cephalopods

Octopus, cuttlefish and squid all evolved into their shell-less forms more recently and their exact lineage is still much debated. Even these most recent cephalopods have nevertheless lived in the ocean for almost 60 million years.

The octopus family has been the subject of several excellent books and studies in the last decade and is variously hailed as including some of the smartest animals on the planet and even "the closest we will come to meeting alien life". These highly varied animals continue to astonish, delight and scare in equal measure. For many, they remain quite the most dystopian animals in the ocean, obviously recalling the fears on which the Kraken legend was based.

I clearly remember our regular dusk dives on a remote reef in Raja Ampat, Indonesia, during a 2001 expedition where we observed the daily emergence of a spectacular new octopus. None of our group had ever encountered anything like it. This was the highly stylised drawing I did from memory in my sketchbook after a dive. Further down the page is a photo taken some years later after our first encounter.

Since then, it has been given a proper, and entirely appropriate, taxonomic name *Wunderpus photogenicus*.

Comparison of a modern *Nautilus* shell and a 200-million-year-old fossil Nautilus – cross-sections showing the chambers, connecting siphuncle and body chamber.

The living *Nautilus pompilius* species together with the rare, and latest evolved, *Allonautilus scrobiculatus*.

Photographed together for the first time in 2015, near Manus, in the Admiralty Islands, Papua New Guinea.

Fossil Nautilus shells from the Jurassic and Cretaceous, 200 to 100 million years old. The one on the left has had the shell polished off to show the chambers

Nautilus: The shelled cephalopod

The Chambered Nautilus is one of the oldest living things on our planet. Since the dawn of civilisation its form has inspired artists, designers and architects. Nautilus has survived whatever the world has thrown at it for more than 500 million years, persisting even as dinosaurs and many other life forms vanished. It is only living externally-shelled cephalopod

Much of what is being written about Nautilus in the media has to do with its potential and imminent extinction – they say it is a 'living fossil', threatened by unscrupulous fishermen, insatiable interior designers and money-grabbing corporations hell bent on mining the deep-sea habitats that are home to Nautilus.

While there may be reason for these concerns, there is an alternate view that, despite having been abused for decades, Nautilus may now be thriving and enjoying a new era of diversification.

Nautilus species may well be thriving and diversifying.

Recently, I provided samples of aragonitic Nautilus shells from my fossil collection to be used for research into nitrogen isotopes contained in them. Initial results appear to establish new insights into the survival secrets of Nautilus. One possible conclusion is this: that the Nautilus family lived in shallow water and were carnivorous right up to the mass extinction at the end of the Cretaceous. At that point, threatened by the high acidity of the oceans, they adapted their lifestyle by searching out deep water and changing their diet to being scavengers. It certainly appears that they haven't eaten any live food since, but are thriving again in waters 400–700 m deep.

We all await the results of these studies and the formal description of any new species with interest.

FIJI Sun February 27, 2019

News Sunbiz Sports World Entertainment Opinion Lifestyle Siga Vou

NATION

US Scientists Find New Marine Species In Savusavu

A new species of nautiluses (an aquatic species) was found last week in Savusavu by scientists from the United States of America. Director of Fisheries Aisake Batibasaga made the remarks

By Swashna Chand , Suva

01 Aug 2018 11:00

From left; Biology Professor Peter Ward of the University of Washington in Seattle, Marine Biology Professor Gregory Barord of the Central Campus Iowa in United States and Fisheries director Aisake Batibasaga in Suva on July 30, 2018. Photo: Swashna Chand

During the past year there have been reports of new species being identified in the Pacific.

Phylum Arthropoda
ANIMALS WITH EXOSKELETONS

Arthropods are invertebrate animals with an external skeleton (exoskeleton) rather than the internal skeletons of the chordates. They have a segmented body with flexible joints. The limbs and body plates are made of chitin. This rigid structure prevents growth, so arthropods need to replace it periodically by moulting.

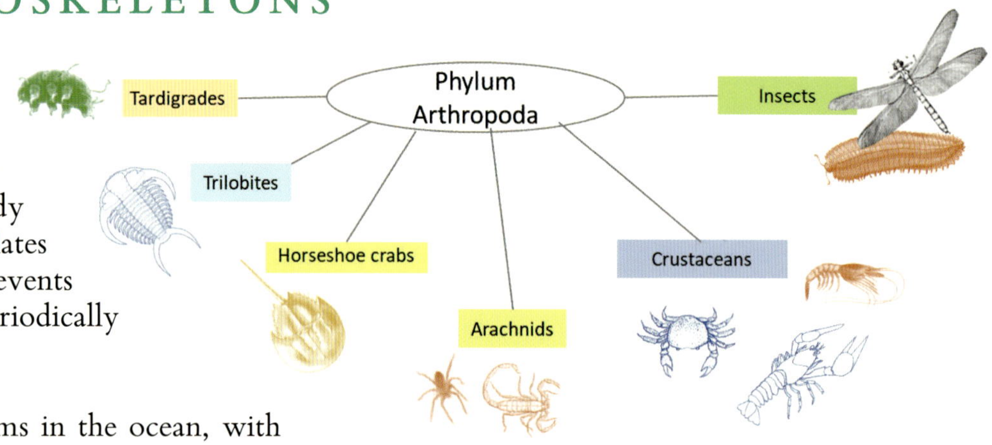

Arthropods provide the earliest evidence of land animals.

They were amongst the earliest life forms in the ocean, with uncertain fossils dating back to before the Cambrian 555 mya. Trilobites, horseshoe crabs and crustaceans remained primarily aquatic, while insects, arachnids., and tardigrades made an early success of escaping the watery world and living on land. Of all these groups, only the trilobites became extinct.

Today, with more than a million described species, arthropods make up almost 80% of all animal species alive on earth – primarily in the form of insects, but they are largely absent from the ocean. Crustaceans, the most common living arthropods in the ocean, are now thought by some researchers to be most closely related to insects.

Today, arthropods are largely absent from the ocean.

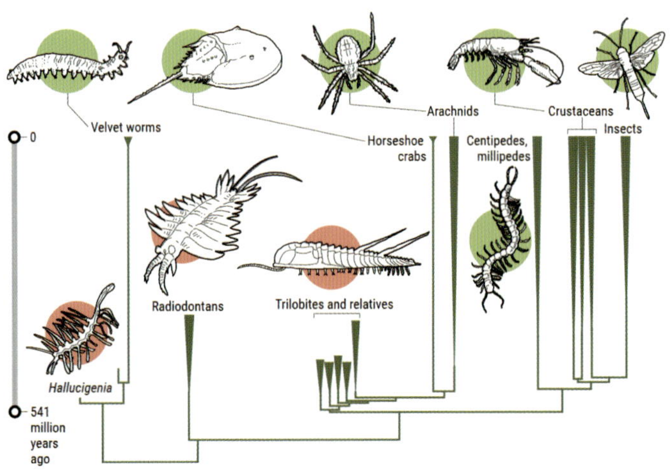

A proposed family tree of arthropods showing the complex relationships between extinct and living groups. The relationships between insects and crustaceans are especially fraught. Source: N. Desai/Science. Data Jo Wolfe/MIT.

A living crab drawn after a dive on an Indian Ocean reef.

A fossil crab *Eriphia cocchi* from the Pliocene (approx. 5 mya), San Miniato, Pisa, Italy. Size130 mm.

Crustaceans : The aquatic arthropods

Barnacles are crustaceans too.

In the ocean, the living arthropods are primarily represented by the crustaceans, including crabs, shrimps, crayfish and, to the surprise of many, barnacles. Yes, despite their appearance, barnacles are crustaceans, not molluscs.

Crustaceans range in size from microscopic to the giant Japanese spider crab with a leg span up to 5.5 m. Many of the smaller crustaceans live in symbiotic relationships on various hosts, and others, such as hermit crabs, live inside discarded shells.

The largest land-living arthropod is the coconut crab *Birgus latro* – it so well adapted to life on land that it would drown in water within a matter of hours. These giant island crabs certainly have a fascination for shiny objects – on the remote Aldabra Atoll I often had to scuttle after them in the dead of night to retrieve parts of my underwater camera rig.

No one has yet agreed where the sea spiders fit into the evolutionary structure and every few years a fresh theory about their true position emerges. It is agreed that they are not true spiders, or arachnids, no matter how much these cryptic creatures resemble them.

Commensal shrimp living on a sea cucumber.

Giant coconut crab *Birgus latro* clambering up palm tree, Aldabra Atoll, Indian Ocean.

A white-spotted hermit crab *Dardanus megistos* living inside a large *Charonia tritonis* triton shell, photographed at night off Pemba, Mozambique.

A sea spider *Nymphon leptocheles* grazing on a hydroid.

A 200 million-year-old fossil and (below) images of a living Atlantic horseshoe crab *Limulus polyphemus*.

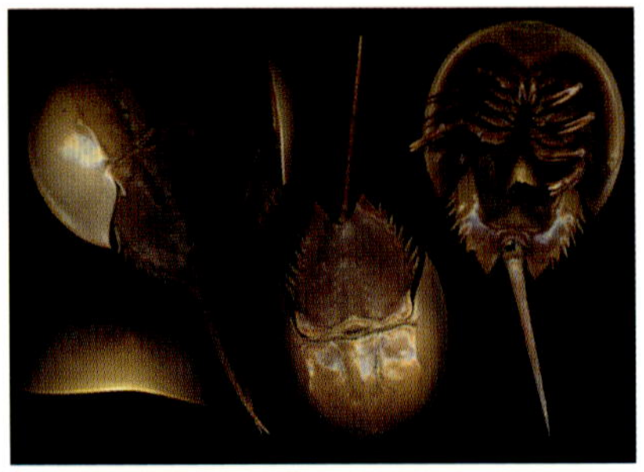

Horseshoe crabs eggs in the palm of a hand.

Horseshoe Crabs

The most constant of the arthropods are certainly the horseshoe crabs. They first appeared in the Ordovician 472–480 million years ago, and the first modern-looking horseshoe crabs emerged in the Triassic about 200 million years ago. Their body plan has changed little, and they have survived all the mass extinctions and several minor ones.

Horseshoe crabs are one of the most enduring lineages on the planet and show very low potential for species radiation. In this sense they are an interesting analogy to the Chambered Nautilus.

Popular perceptions of horseshoe crabs are based on the annual spawning events that occur along the east coast of America, where mating adults are seen in their thousands crawling up on beaches to deposit and fertilize their eggs, the exact moment in their life when they are at their least elegant.

Horseshoe crabs have evolved incredible powers of healing and rejuvenation when injured, largely because they live in such hostile coastal waters and must cope with teeming bacteria, anoxic environments, extreme weather, high temperature fluctuations plus sudden floods of fresh water. Adult animals have survived being frozen in ice and being 'boiled' in water up to 460°C and are able to transition between marine, estuarine and freshwater habitats with ease.

The body of a horseshoe crab comprises a deep vaulted carapace which protects the feeding apparatus and appendages. They famously have ten visual organs, bright blue blood and a highly effective immune system, research into which led to the discovery of limulin, effective in the detection of bacterial endotoxins.

Tardigrades

When fully grown, tardigrades are smaller than this full stop. You will, inadvertently, have consumed many.

At the extreme end of the arthropod spectrum are the creatures known informally as *water bears* and *moss piglets*. Tardigrades are not known for their appealing looks, but they are the ultimate survivors, not fazed by extreme temperatures, pressures or radiation, and they can live for long periods without air, water or food.

Tardigrades are water-dwelling segmented animals that are found literally everywhere, from the deepest ocean to mountain tops, and in the most hostile environments known – from mud volcanoes and tropical rain forests to the Antarctic – and anywhere in your garden. They are so small that a few can lie comfortably on a pinhead. We inadvertently consume them alive and often – on fresh salads and vegetables or in water.

Tardigrades are the most resilient animals known. They can survive extreme conditions that would be instantly fatal to nearly all other known life forms, except some bacteria.

When a tardigrade dries out it sheds almost all the water in its body and curls up into a dry husk. Its metabolism slows to 0.01% of normal, entering suspended animation that resembles death. It can stay in this state for decades, only reanimating when it encounters water. Pictures show the tardigrade *Paramacrobiotus richtersi* alive, partially desiccated and in its shrivelled 'tun' state, at right.

There are about 1,150 known living species and the oldest fossils date from 530 million years ago, in the Cambrian period.

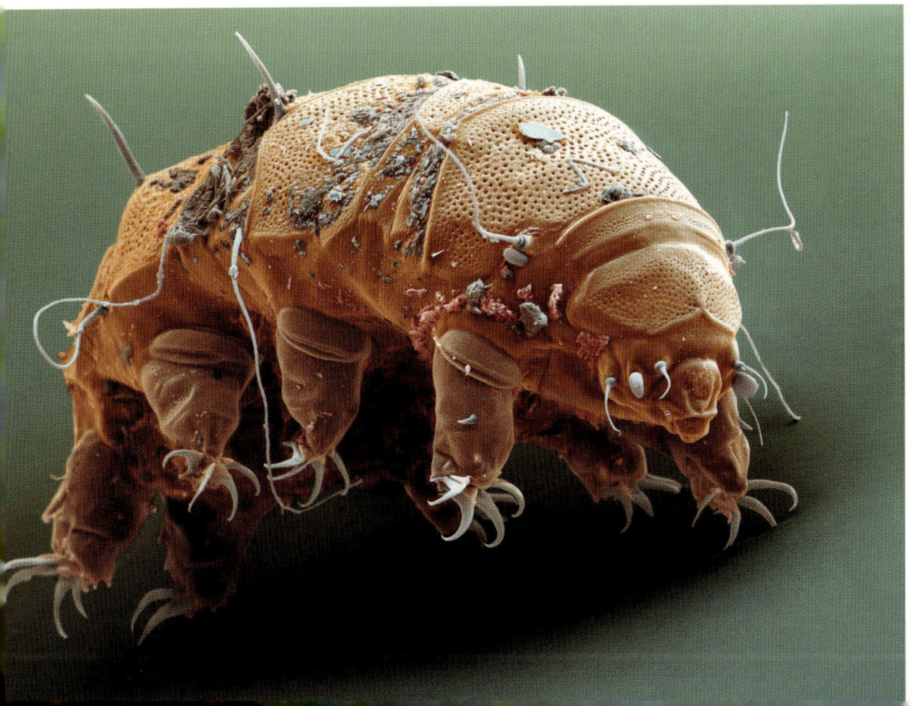

Above is a fossil fish *Heterolepidotus ornatus* from the Triassic of Austria 220 mya. The background image shows a mixed school of barracudas, surgeonfish and others in the open ocean off Assumption Island, near Aldabra in the Indian Ocean.

Phylum Chordata
FISH

For the first 150 million years of complex life, fish were the pinnacle of evolutionary achievement.

John Long in *The Rise of Fishes*.

Today, fish are the most diverse form of vertebrates with more than 23,000 identified species. They were amongst the very first animals with backbones and are the ancestors of all reptiles, birds and mammals – including of course, us.

The term *fish* however is severely misused. Barracudas, sharks, goldfish, dolphins are all 'fish' in the sense that they all have 'fish' characteristics, but they are not closely related to each other in evolutionary terms. Present-day dolphins and Jurassic ichthyosaurs evolved very similar 'fish shapes' – but one is a mammal and the other a reptile. In biology, the term fish is strictly used to describe any animal with a backbone that has gills throughout life and has limbs, if any, in the shape of fins.

This *Pikaia* fossil was discovered in 1911 in Yoho National Park in the Rocky Mountains. 505 mya.

If you look at the pull-out Phylogeny Chart you will see that a primitive skull and backbone emerged with the chordates 530 million years ago. The first primitive 'fish-like' specimens from the Burgess Shale, such as the early chordate *Pikaia* sp., just had a flexible rod of cartilage (a *notochord*) for its skeleton. Superficially they were quite *fishy,* but perhaps not quite a fish yet.

Life reconstruction of *Pikaia gracilens*.

The four major groups of fish evolved from these from the Ordovician to the Devonian Periods:

- **Primitive Jawed Fish** These are extinct – an example is the Placoderm we met earlier.
- **Cartilaginous Fish** These include sharks and rays.
- **Bony/Lobe-Finned Fish** The root stock from which all four-legged vertebrates evolved. These include the living coelacanth.
- **Bony/Ray-Finned Fish:** literally all the other true fish, they range from large pelagic fish to the tiniest reef critters – from schooling barracudas and surgeonfish (on the facing page) to the charming sea goldies *Pseudanthias* sp. (at right).

A fossil bony fish
Xiphactinus audax Cretaceous, 83 mya.
Kansas, USA.

A living blacksaddle rockcod
Plectropomus laevis
Aldabra Atoll, Indian Ocean.

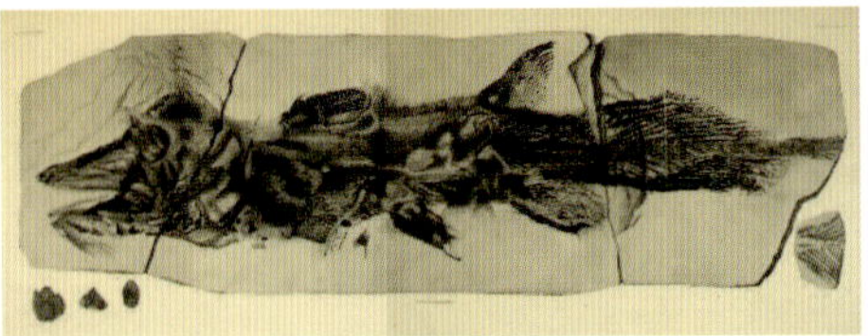

Fossil coelacanth – from Fossil Fishes and Fossil Plants of the Triassic Rocks – J Newberry 1888.

Although I had come prepared, that first sight hit me like a white-hot blast and made me feel shaky and queer, my body tingled. I stood as if stricken to stone. Yes, there was not a shadow of doubt, scale by scale, bone by bone, fin by fin, it was a true Coelacanth.

JLB Smith, in The Search Beneath the Sea

Lobe-finned Fish & the Coelacanth

One of the original lobe-finned fish from the Silurian and Devonian is still alive today, almost unchanged – the coelacanth. The first evidence of a living coelacanth was a *strange* fish caught by a Russian trawler in the Mozambique channel and identified by JLB Smith as a coelacanth after the ship docked at East London harbour in South Africa's Eastern Cape Province in 1938.

Old Four Legs – The Dino-Fish: JLB Smith a few moments after the identification of the second coelacanth in the Comoros Islands in the Indian Ocean, 29 December 1952.

The scientific importance of the discovery of the coelacanth cannot be overestimated and points to the origins of all vertebrate animals, us included. Embryos, including our own, exhibit gill slits and tails at certain stages of development, again pointing to our true origins. JLB Smith did not call the coelacanth *Old Four-Legs* flippantly. The lobefins of the coelacanth indicate a transitional move from marine to land-dwelling animals. Coelacanths are so primitive that they are more closely related to reptiles and mammals than to the common ray-finned fishes.

A Comoros coelacanth at depth, photographed from the Jago submersible.

The coelacanth quickly became known as the *Dino-Fish,* but it wasn't until 1986 that the first living coelacanth population was filmed and studied in the Comoros Islands. A 20-year study by Hans Fricke of the Max-Planck Institute in Germany together with the JLB Smith Institute of Oceanography, on whose Management Board I served, finally established that this was a stable population of about 500 individuals. Hans Fricke's iconic little yellow submarine *Jago* was for years the visible symbol of research into the coelacanth.

Hans Fricke and his Jago submersible.

Indonesian coelacanth photographed off Sulawesi, Indonesia.

Several more populations were subsequently found: in 1997 off Sulawesi in Indonesia; in 2000 at Sodwana Bay in South Africa and in 2007 off Tanzania – all after the coelacanth was thought to have been extinct for 50 million years. If you're a really adventurous diver with unlimited resources, then you can arrange to swim with this relic lobe-finned fish today.

Manado Tua, a volcanic island in the Celebes Sea off the north east coast of Sulawesi, where the second living population of coelacanth was discovered.

Bony/Ray-finned Fish

These are fish that have skeletons of bone tissue, as opposed to cartilage such as sharks and rays. The earliest fossils date back to 420 million years ago and are followed by an incredibly diverse fossil record that speaks of constant evolution, diversification and adaptation to almost all marine environments. Amongst them are today's 'sea dragons', the common name for the relatives of sea horses.

Fish have been a constant primary food source for us and are increasingly under pressure from growing populations and environmental degradation. Fish farms and selective breeding address some of the culinary demands.

Other fish, such as the decorative Siamese fighting fish are now directly dependent on human intervention for their survival. These fish have evolved from oceanic salt-water ancestors to live exclusively in standing stagnant waters with low oxygen content in canals, rice paddies and floodplains around the Mekong basin of southern Asia. In the wild they are dull, with short fins, and exhibit strong colours only when agitated. Thankfully, they have been highly prized for centuries and subject to extensive selective breeding to accentuate their vibrant colours and extravagant tails. In their natural environment they are listed as *Vulnerable* by the IUCN. Their future survival, and certainly their future evolution, is directly dependent on human intervention and their immense popularity in the aquarium trade.

Two intertwined Siamese fighting fish *Betta splendens*.

A spectacular modern 'Sea Dragon', the Ornate Ghost Pipefish *Solenostomus paradoxus*, photographed in Papua Province, Indonesia and a quick drawing from my sketchbook.

Fossil batfish *Eoplatax papillo* found at Monte Bolca, Italy.
Eocene 56–34 million years ago.

When the European and African plates collided to form the modern-day Alps between 24 and 50 million years ago, the Tethys Ocean and the fossils forming on its floor were uplifted. One superb locality in which these fossils are preserved is in the area of Monte Bolca in Italy. The environment in which they were fossilised was low in oxygen, preventing both decay and damage by scavengers. Preservation is so good that even skin colour can sometimes be determined. These fossils have been known since the 16th century.

A school of batfish cruising at the edge of the East Channel of the Aldabra Atoll, Indian Ocean. 20th century.

In the 1950s, Jacques Cousteau wrote about his experiences drift-diving in the channels of Aldabra, the second largest coral atoll in the world at 34 km long by 13 km wide. It has a large shallow lagoon which largely empties twice daily during low tide dragging its marine population into the welcoming arms of predators out at sea.

At right is a juvenile batfish, showing the dramatic shape and colour changes they undergo as they mature.

The giant moray *Gymnothorax javanicus* can be up to 3 m in length and weigh up to 30 kg. I have experienced them up close in the Red Sea and felt their silky skin against mine as we slid past each other under crevaces – closeness was always at their initiative, not mine.

A pair of Dark-spotted moray eels *Gymnothorax fimbriatus* in typical posture. Milne Bay, Papua New Guinea.

The Dragon *Moray Enchelycore* pardalis, could easily be a mythical dragon with its narrow jaws, vivid colours, long tubular nostrils and glass teeth. It is seldom seen, and then mostly at night when it hunts. This photo is of the first specimen to be recorded this far west in the Indian Ocean, around the Comoros Islands in 1997.

Moray eels are visually the closest living things to the *Sea Dragons* of ancient seas. Although significantly smaller in stature, divers are usually able to get close enough to be suitably arrested by their prominent fangs. Their typically gaping mouths are not a sign of aggression but a result of the position of their gills located far back in their mouths. This requires them to keep their mouths open to facilitate breathing. Morays also have large teeth and relatively poor eyesight as they lie in wait in holes and crevices in the reef to ambush prey. I have often been bitten by smaller morays that must have mistaken my fingers for an octopus, squid or cuttlefish – amongst their favourite food – as I was photographing critters in and around their favourite 'holes'.

The lair of a large honeycomb moray *Gymnothorax favagineus* under an overhang is shared with a cleaner shrimp *Lysmata grabhami* and an unidentified red crab. Chain Reef, Sodwana Bay, South Africa.

Sharks and rays are amongst the earliest fish, dating from the start of the Devonian Period 410 mya. The Golden Age of Sharks was during the Carboniferous, and the Modern Sharks emerged in a profusion of forms during the Jurassic and continued to evolve along separate paths for nearly 200 million years. This sand tiger shark encountered on one of our dives has similar features to the *Cladoselache* sharks from the Devonian 380 mya. Picture insets at top show a fossil and an artist's reconstruction.

Sharks

Today's shark population is easily as diverse as it has ever been, with almost 500 described species. There are the large laid-back plankton feeders, such as the basking shark and the whale shark (actually the world's biggest fish), hundreds of small and mid-size sharks, and, at the other extreme, the fast, agile and mean predators so loved by pulp fiction and epitomised by the great white shark.

The shark *Carcharias taurus* is known by many names around the world – the 'raggie' or 'spotted raggedtooth shark' in South Africa, the 'sand tiger shark' in the USA and New Zealand and the 'grey nurse shark' in Australia. It is not related to the 'tiger shark'. Its proper name *Carcharias taurus* clears up that confusion, It is a close relative of the great white shark *Carcharodon carcharias*.

The sand tiger shark is one of the larger shark species, and it is feared more for the terrifying set of teeth dangling out of its mouth than for its bite. Once a year, pregnant sand tiger sharks gather around a small rocky reef off the east coast of South Africa. We headed out to the dive site on a small inflatable after some rough weather, geared up and dropped down onto the shallow reef. Visibility was poor with persistent surges churning up the sandy bottom. When the limited fog of our vision cleared, we saw a number of large females drifting in and out of the gloom.

We had dived with pregnant sand tiger sharks before and knew them to be shy and timid if not provoked. Being amongst these serene mothers-to-be was always a profound privilege as they went about their gestation rituals.

The sand tiger shark illustrates several common and definitive evolutionary traits common to many sharks. Its jaws contain at least five concentric rows of teeth, embedded in connective tissue stretched over a cartilage jaw. Each of the inner teeth is ever moving forward to replace older teeth lost at the front – many thousand teeth in a typical lifetime of around 15 years.

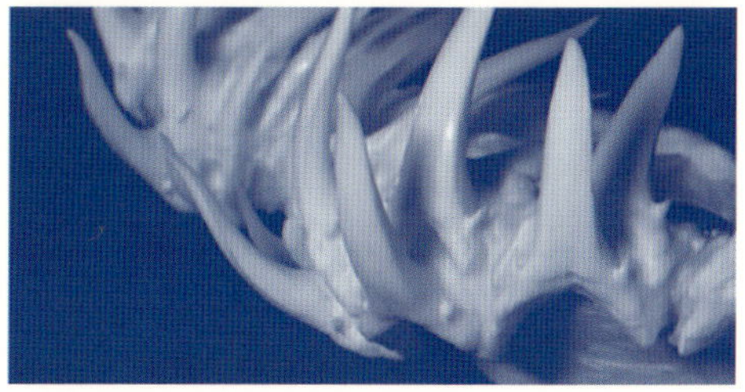

Wobbegong or carpet shark *Eucrossorhinus dasypogon* resting on a large plate coral. Raja Ampat, Papua Province, Indonesia.

Catsharks are distinguished by their elongated cat-like eyes. They are poor swimmers and nocturnal and tend to languish under rock overhangs or in caves during the day. Aldabra Atoll, Indian Ocean.

The reproductive strategy of the sand tiger shark is much more *unusual*. Each mother produces masses of eggs (more than twenty thousand were found in one shark), packaged in tiny parcels of about ten, making hundreds of small egg bags in each of her two uteri. The embryos that develop first have an unfair advantage over the others and use the other egg parcels as a food supply, consuming both the unfertilised eggs and the fertilised embryos. The embryonic sand tiger shark literally consumes its own siblings – until only one shark is born from each uterus. This process of intra-uterine cannibalism is called *oophagy* and represents natural selection with a vengeance – even before birth. Oophagy is thought to occur in all sharks in the order Lamniformes – the mackerel sharks.

At birth the sand tiger sharks are each about a metre long and weigh six kilograms. At first limp and unable to swim, they take a few hours to tone up their muscles and stiffen the fins, and then they fend for themselves immediately. This practice may lead to larger embryos and help prepare offspring for a predatory lifestyle. The mothers show no parental care, and, starved from months of abstinence, leave the area immediately after pupping to avoid feeding on their own offspring.

Some of today's shark species also don't do too badly in the *weird* stakes. Take the carpet sharks in the family Orectolobidae. They are found chiefly around Australia and Indonesia and are sometimes called *wobbegong sharks* – from an Australian Aboriginal word meaning "shaggy beard", referring to the growths that surround their mouth.

In my experience they are amongst the most sedentary of all sharks and are mostly found resting on the sea floor and on large plate corals. The shaggy beards help with their magnificent camouflage and support their role as ambush predators. Typically, they attack by lunging upward and grabbing fish. Their teeth, though small, are sharp and have been known to inflict injury on humans.

Rays

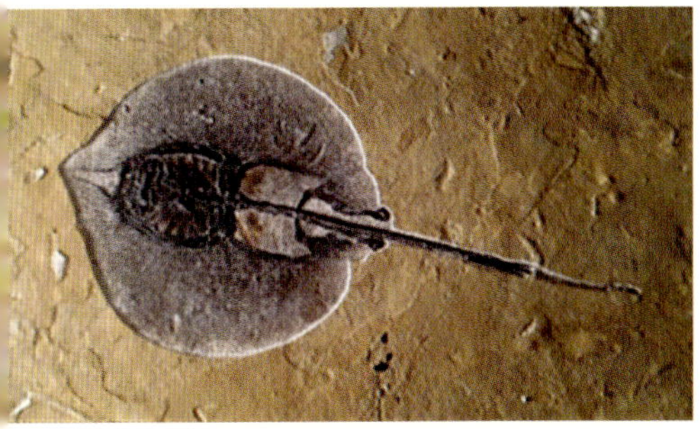

Ray fossil from the Clear Creek Quarry in the Green River Formation of Wyoming. Eocene 48–52 mya.

Rays are cartilaginous fishes related to sharks, within a family called Elasmobranchs that also includes sawfish and skates. They evolved their highly flattened shapes during the late Jurassic, so represent a continuity of more than 150 million years.

They live primarily on the sea floor in coastal areas, although some are known to live in deep water beyond 3,000 m. Many have venomous tail spines. They are strong swimmers and we have witnessed eagle rays swimming leisurely against a ten-knot current in the tidal channels of Aldabra in the Indian Ocean.

The bluespotted ribbontail ray *Taeniura lymma* is a species of stingray that excavates sand pits in search of molluscs, worms, shrimps, and crabs. Sodwana Bay, South Africa.

The giant manta rays by comparison are open ocean dwellers and can reach wing spans of seven meters across. They are amongst the most graceful large creatures in the ocean. In the waters off Misool, Papua Province of Indonesia, we encountered both the white and black colour morphs of *Manta birostris*. They were long thought to be different species, but this was disproved by DNA studies in 2001.

White and black morphs of the manta ray *Manta birostris*. Raja Ampat, Indonesia.

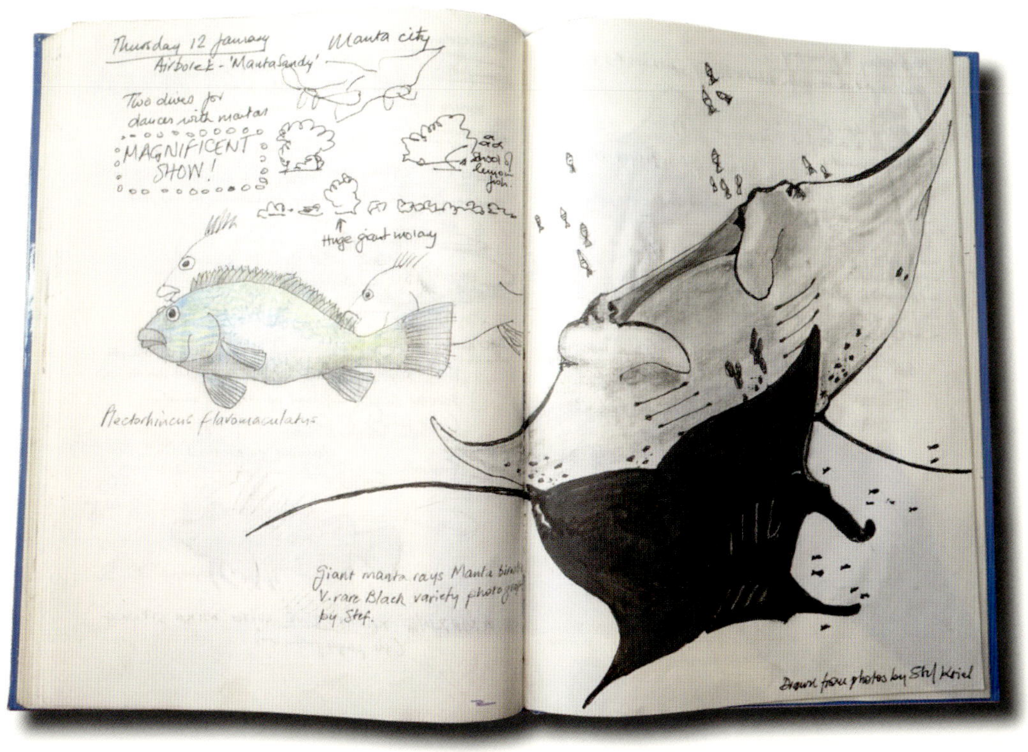

Phylum Chordata

CLASS REPTILIA

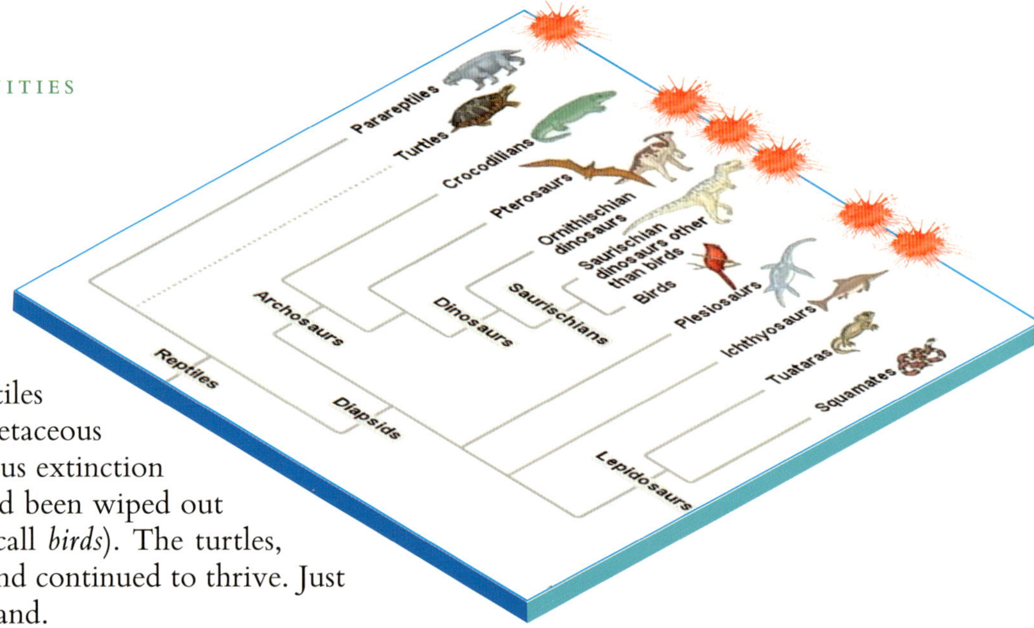

Primitive para-reptiles first arose in the Carboniferous Period, and large reptiles dominated life on land in the Jurassic and Cretaceous Periods 200–66 mya. By the end-Cretaceous extinction most of the large reptiles and dinosaurs had been wiped out (except the avian theropods – which we call *birds*). The turtles, crocodilians, and the squamates survived and continued to thrive. Just two species of tuataras survive in New Zealand.

As you can see from the chart above, reptiles do not all have common recent ancestors. For example, crocodiles are more closely related to birds than they are to lizards. For this reason, many scientists prefer to include birds as part of Reptilia as well, thereby making it a monophyletic class with a common ancestor.

Amongst the ancient reptile lineages, only the turtles, crocodilians, squamates (including the lizards and snakes) and the birds survive today.

The marine iguana *Amblyrhynchus cristatus*, Galapagos Islands.

Squamates are now the largest order of living reptiles. They include lizards and snakes, animals from the smallest dwarf gecko at 16 mm to the largest tropical snakes at more than 9 m in size.

Reptiles are today much more common on land than in the ocean.

Few reptiles are now living in the ocean: just some species of turtles, salt water crocodiles, sea snakes and a solitary species of marine iguana in the Galapagos Islands.

Turtles harvested for their meat on a remote Indian Ocean island, the precious 'tortoise shell' left to crack and bleach in the sun.

Aldabrachelys gigantean the Aldabra giant tortoise in its natural habitat on Aldabra

The shell of the world's smallest land tortoise *Homopus signatus* just 75 mm long – atop the shell of the world's largest at more than 1 m.

Turtles

Turtle is an umbrella term for all reptiles in the order Testudines – these include turtles, tortoises and terrapins. Tortoises live exclusively on land, while turtles and terrapins live in the water nearly all the time and only come ashore to lay eggs. While in North America the above terminology is common, in Great Britain, the word turtle is used for sea-dwelling species, but almost never for tortoises. What turtles (and tortoises!) all share is a shell to which the ribs and vertebrae are fused.

The first true sea turtles evolved over 110 mya during the Cretaceous period. There is now evidence that all modern turtles are descended from a terrestrial ancestor, a reptile who lived before most of the dinosaurs. Of course, even the sea-dwelling turtles still return to the land to lay their eggs so cannot be regarded as true ocean-dwellers. Their young hatch and return to the sea where they spend their entire life – then go full-circle to lay their eggs back on land years later.

Tortoises are generally regarded as vegetarians, while other turtles are thought to be omnivorous. But that is not always the case. I have personally witnessed the giant tortoise *Aldabrachelys gigantean* feasting on a deceased family member in their home environment, the Aldabra Atoll in the Indian Ocean. It has the largest population of giant tortoises in the world, numbering more than 100,000 – five times the number that inhabit the Galapagos islands whose numbers dipped as low as 3,000 in the 1970s.

Although the demand for *tortoise-shell* (actually the shells of the hawksbill turtle) has been largely eradicated due to changing fashions, legislation and artificial alternatives, turtles are still endangered by human pressure on the beaches where they lay their eggs and in remote areas where they are still taken as food.

A large fossil female sea turtle *Alienochelys selloumi* found in the phosphate mines in the Ouled Abdoun Basin in Morocco. It dates from the end of the Cretaceous Period about 70 mya The carapace itself is longer than 2 m and the skeleton is 3 m long from head to impressive tail (shown in top picture). *Alienochelys* is an ancestor of the modern leatherback turtle *Dermochelys coriacea* (below) which can reach a similar size and its front flippers can grow up to 2.7 m long.

Squamates: The scaled reptiles

Squamates are now the largest order of living reptiles, which includes lizards and snakes, diverse animals from the smallest dwarf gecko at 16 mm to the largest tropical snakes at more than 9 m in size. Few of this reptile group live in the ocean. The marine iguanas and sea snakes are the exceptions. Sea snakes are almost independent of land and are often found far from shore.

Most species of sea snake give birth to live young in the water, just like the ichthyosaurs before them. One of the few sea snakes that lays eggs on land is the yellow-lipped sea krait *Laticauda colubrina*, also known as the banded sea snake, found in the tropical Indo-Pacific Ocean. It spends much of its time underwater in order to hunt, but returns to land to digest, rest, and reproduce.

Their venom is a very powerful neurotoxic protein, fatal to humans. A close encounter with sea snakes is always a humbling adrenalin-rush – as was this extended meeting in Raja Ampat, Papua Province, Indonesia.

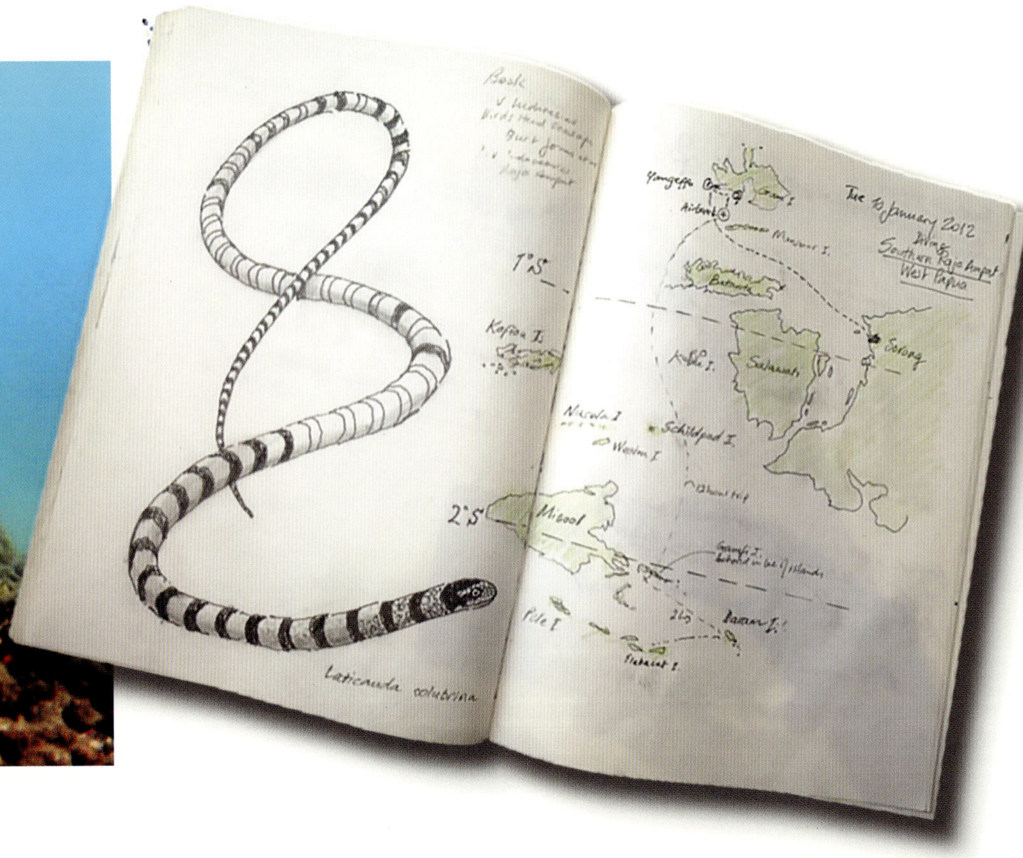

Phylum Chordata
CLASS MAMMALIA

The origins of marine mammals

The dramatic destruction of plant life on land around the end of the Cretaceous Period must have made it seriously attractive for mammals already living at the edge of rivers and shores to venture into water searching for their next meal. The extinction of the dominant marine reptiles, the ichthyosaurs, plesiosaurs, pliosaurs and mosasaurs, led to a super-abundance of fish, making the seas even more attractive.

By 50 mya, creatures such as *Pakicetus*, with crocodile-like skulls, were likely moving around the edges of lakes and rivers, and in the water doing what some have described as a 'doggie paddle' using their modified webbed feet. Others were long-snouted and otter-like. They all lived in near-shore environments, from saltwater marshes to shallow seas and began a diet of marine animals. Over many generations they developed streamlined features suited to the marine life. Feet and tails broadened to form wide flukes for aquatic propulsion. Fur was almost completely lost.

Pakicetus recreation in the Canadian Museum of Nature.

Manatees & Dugongs

Sirenia (the name comes from the *sirens* of Greek mythology) is the order which includes *sea cows* (manatees and dugong) and their extinct relatives. They are the only living herbivorous mammals to have become completely aquatic.

They attained diversity during the Oligocene and Miocene, from 33 to 5 mya, but have declined as a result of changing climate and environmental pressures from humans.

The skull and teeth of one of the earliest known whale, *Archaeoceti* sp., Upper Coastal Plain of eastern Georgia, Early Eocene 40 mya.

Early fossil whale skull and teeth, from Morocco.

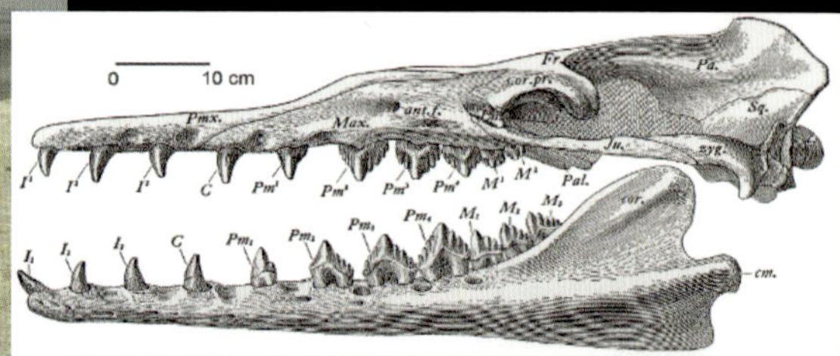

Early fossil whales

We explored the Pisco Formation in the deserts of Peru which contain hundreds of well-preserved fossil baleen whale specimens spread across several hundred square kilometres of a dry desert, at a distance of more than 20 km from the Pacific Ocean. These fossils are ten to twelve million years old.

Whales & Dolphins

By thirty million years ago, some mammals were already completely amphibious. Thought to be the direct ancestors of modern whales and dolphins, *Cetacea archaeoceti* (the "ancient whales") appeared. They represent important steps toward a fully marine existence. These early whales were already very varied, from small dolphin-like animals to gigantic eel-like creatures 15 m in length.

Many of the original forms eventually died out, but not before giving rise to the two groups of whales alive today, the toothed whales and the baleen whales. They include the Yangtze River dolphin, the gigantic blue whales and the small toothed whales we call dolphins. All marine mammals continue to give birth to live young, in the water and still breathe air, not having developed anything like gills.

I have experienced nothing more dramatic than the sight of a large whale teasing itself graciously out of the water or a school of spinner dolphins performing their aerial ballet for our privileged audience far out at sea. Few of us consider that these glorious mammals have thrived in the ocean for more than a hundred times the span of human existence – a continuity of more than 50 million years.

A pod of Gray's spinner dolphins *Stenella longirostris longirostris* photographed in mid-ocean off Madagascar in the Indian Ocean.

THE BIGGEST SPLASH

The strange history of the ocean's biggest animals

For the past 300 million years, the biggest animals in the ocean, all of them except for the sharks and the giant squid, evolved from life forms that lived on land.

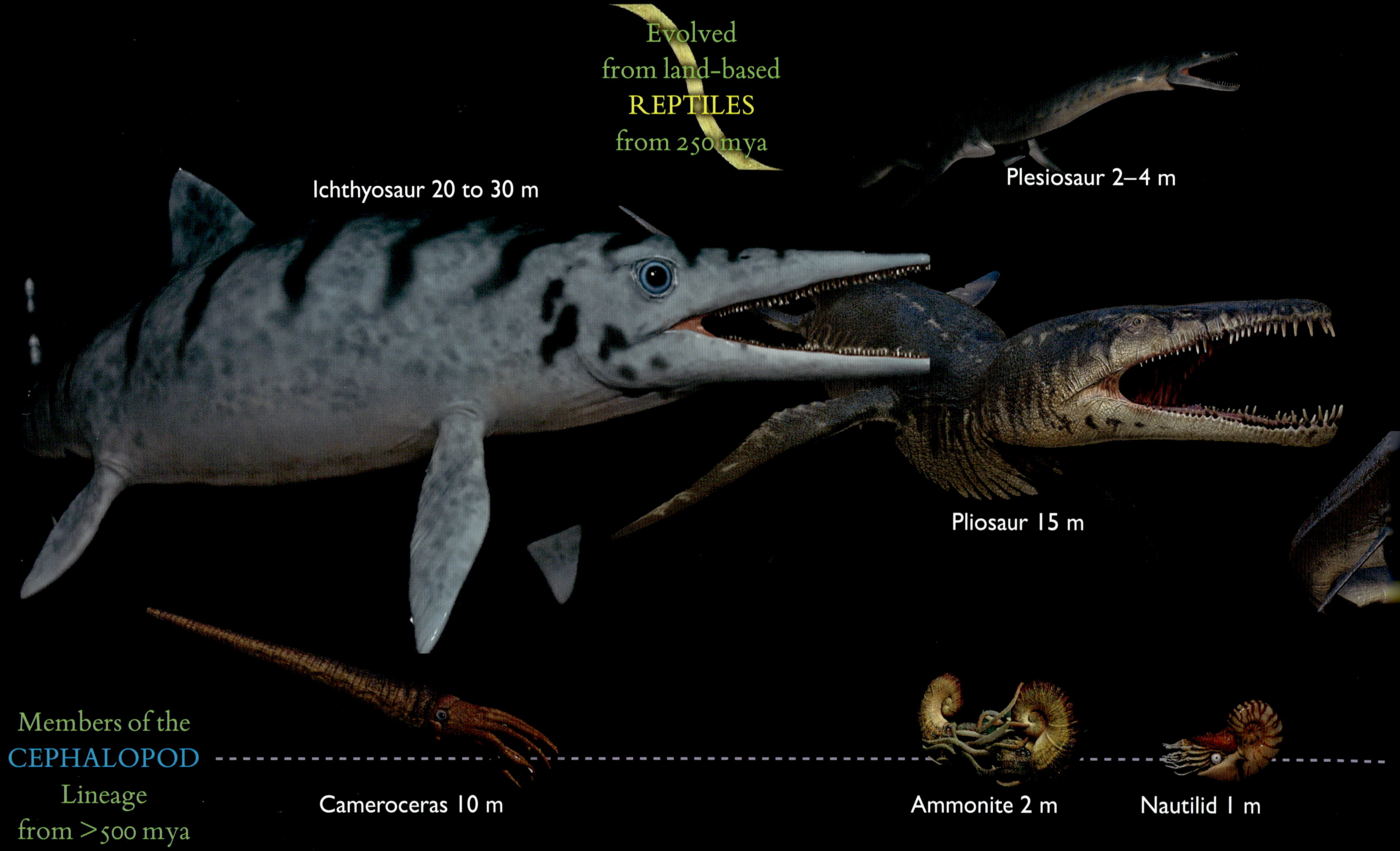

Cephalopods were the largest predators in the ocean until the arrival of the ichthyosaurs

What was it that land-based animals learned that made them so successful in the ocean?

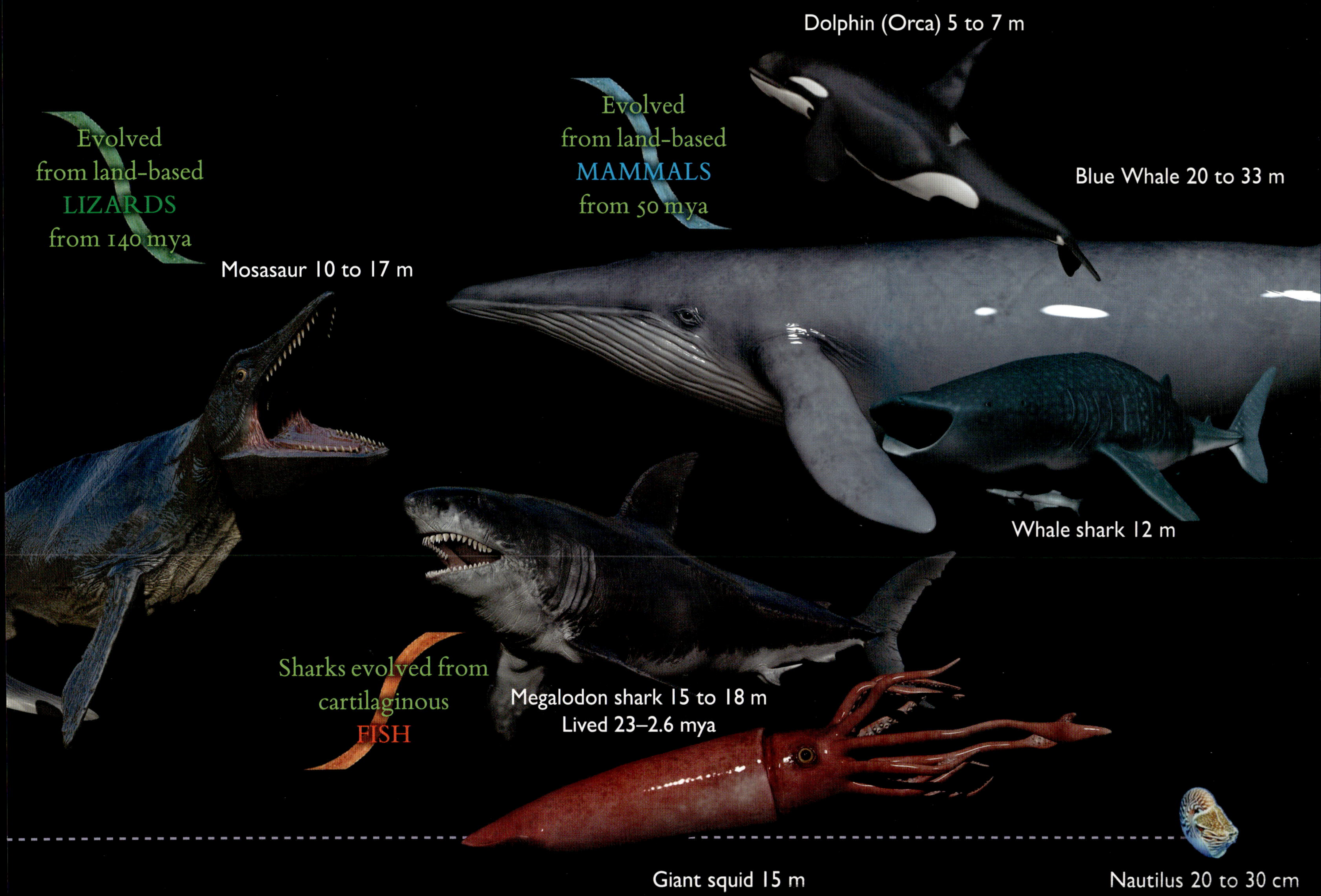

…today's giant squid (also cephalopods) remain one of the largest animals in the ocean.

The timeline of the ocean's biggest animals reveals some interesting facts, and a fascinating continuity, covering hundreds of millions of years. The continuity in this case is the enduring niche for the ocean's biggest animals. For the last 300 million years this niche in the ocean has been filled by evolutionary interlopers from the land. Each one from a different group of land-based animals.

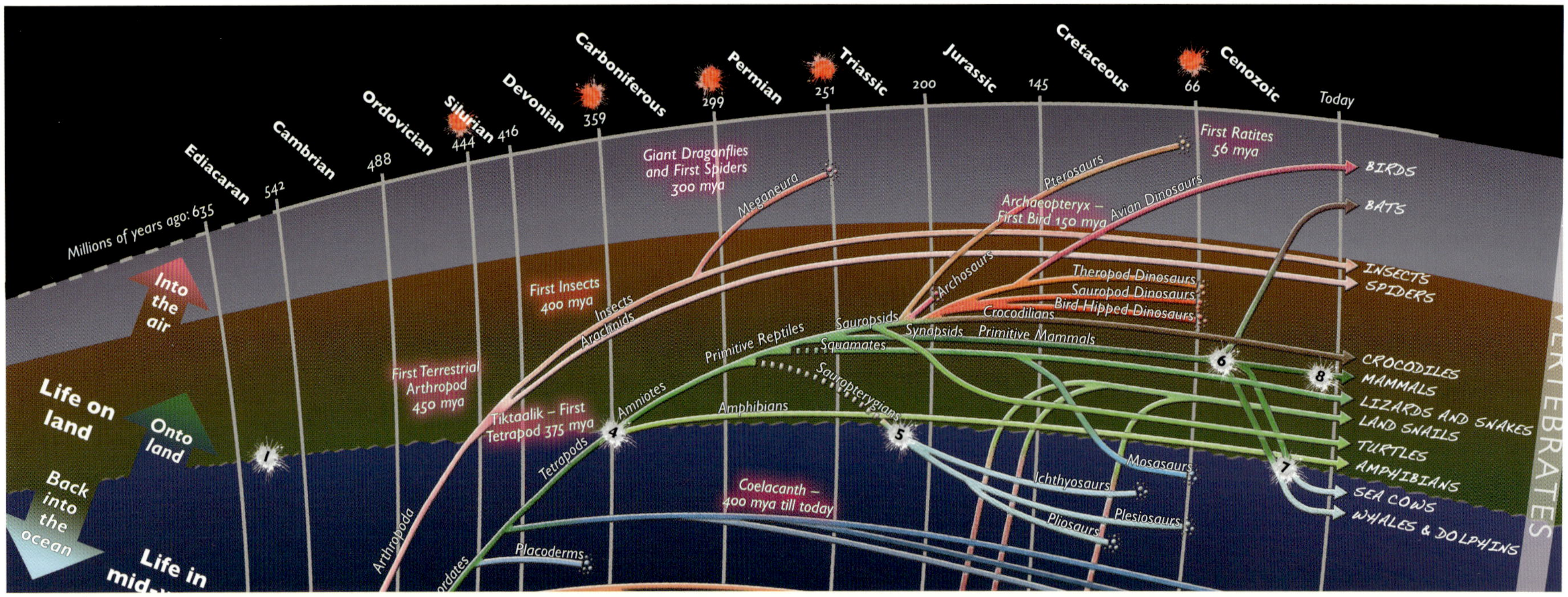

Until the Triassic, the biggest creature in the sea was a cephalopod.

Until the Triassic, the biggest creature in the sea was a cephalopod – the giant *Cameroceras* nautiloid that grew a shell to 10 m long or more. Add to that its body and tentacles and we are really dealing with a substantial animal – perhaps even with a voracious appetite to feed such bulk.

Larger ocean giants, and its most ferocious predators, did not evolve until much later in the Triassic and, perhaps surprisingly, they all originated from groups that previously lived on land.

First it was reptiles that became attracted to life in the sea. Ichthyosaurs and plesiosaurs evolved from a group of unidentified land reptiles around 250 mya. Clear transitional forms have not yet been found: the earliest known fossils of the ichthyosaur lineage are already fully aquatic.

Mosasaurs are believed to have evolved from terrestrial monitor lizards in the Early Cretaceous. With the extinction of the ichthyosaurs and pliosaurs, mosasaurs became the dominant marine predators, but the large ones did not appear until 3 million years after the extinction of the ichthyosaurs.

Whales and dolphins, derived from mammals living on the edge of rivers and coasts, have reached sizes comparable to the largest dinosaurs. Blue whales can grow to 33 m in length, similar to the largest Cretaceous dinosaurs.

Many of the land-based animals that returned to the ocean evolved similar streamlined fish-shaped bodies – an excellent example of convergent evolution. All of these animals also retained the ability to give birth to live young.

After life moved from the ocean onto land, the largest creatures in the ocean evolved from life on land.

These super-predators raise many questions: Was it the super-abundance of fish that enticed these reptiles and mammals back into the sea? What did they learn on land that would make them so successful in the ocean?

Plesiosaurus battling *Temnodontosaurus* (Oligostinus), frontispiece from *Book of the Great Sea-Dragons* by Thomas Hawkins.

While cephalopods were the largest predators in the ocean until the arrival of the ichthyosaurs and plesiosaurs, today's giant squid at 21 m long still remains one of the largest ocean predators. The large sharks almost pale into insignificance alongside these real ocean giants.

It is interesting to reflect on how much we have learned since Thomas Hawkins, an English fossil collector and dealer, first coined the term *Sea Dragons* in his 1840 *Book of the Great Sea-Dragons*. This frontispiece to his book was typical of the fantastical animal scenes that were generated from so little fossil evidence at the time. Many examples from Hawkins' fossil collection can still be found on display at the Natural History Museum in London.

ORIGINS & DESTINIES

Perspectives on life, our role in it and the luxury of choice

If you stand on a beach and stare out at the horizon you can see about 3 miles out...if it's clear you might be looking at 10–20 square miles of ocean. But the ocean is actually 10 million times the size of your view...and on average there are more than two miles of water beneath. The most extreme thing about the ocean is its sheer, inconceivable size.

S. and A. Palumbi *The Extreme Life of the Sea*

We call it *The Blue Planet* for a reason. The NASA image of the Earth on the facing page was taken over the Pacific Ocean and really accentuates that fact – there is very little land in sight. When the crew of Apollo 17 beamed home their first image of the earth on 7 December 1972 (the one at right), our planet was immediately dubbed *The Blue Marble*. This has become one of the most reproduced images in human history.

During thousands of hours spent scuba diving, I have often been overwhelmed by the diversity of life teeming around me and sometimes the enormous depth of the open water beneath me. Gazing across a colourful virgin reef, I have reflected on how similar to a Jurassic reef this all might be. I have experienced the white-hot fear of being left alone in the open ocean, floating in turbulent waves, out of sight of the only boat within hundreds of miles, with dusk approaching. At the worst of times the ocean can be hard and heartless. But, even at the best of times, we cannot conceive the ocean as being anything but boundless and permanent.

For me, that truism was shattered dramatically by the image shown on the following page. Please don't turn over now. Take a moment to imagine that you could take all the water in the ocean and wrap it up into a ball. Close your eyes and think about how big that ball of water might be?

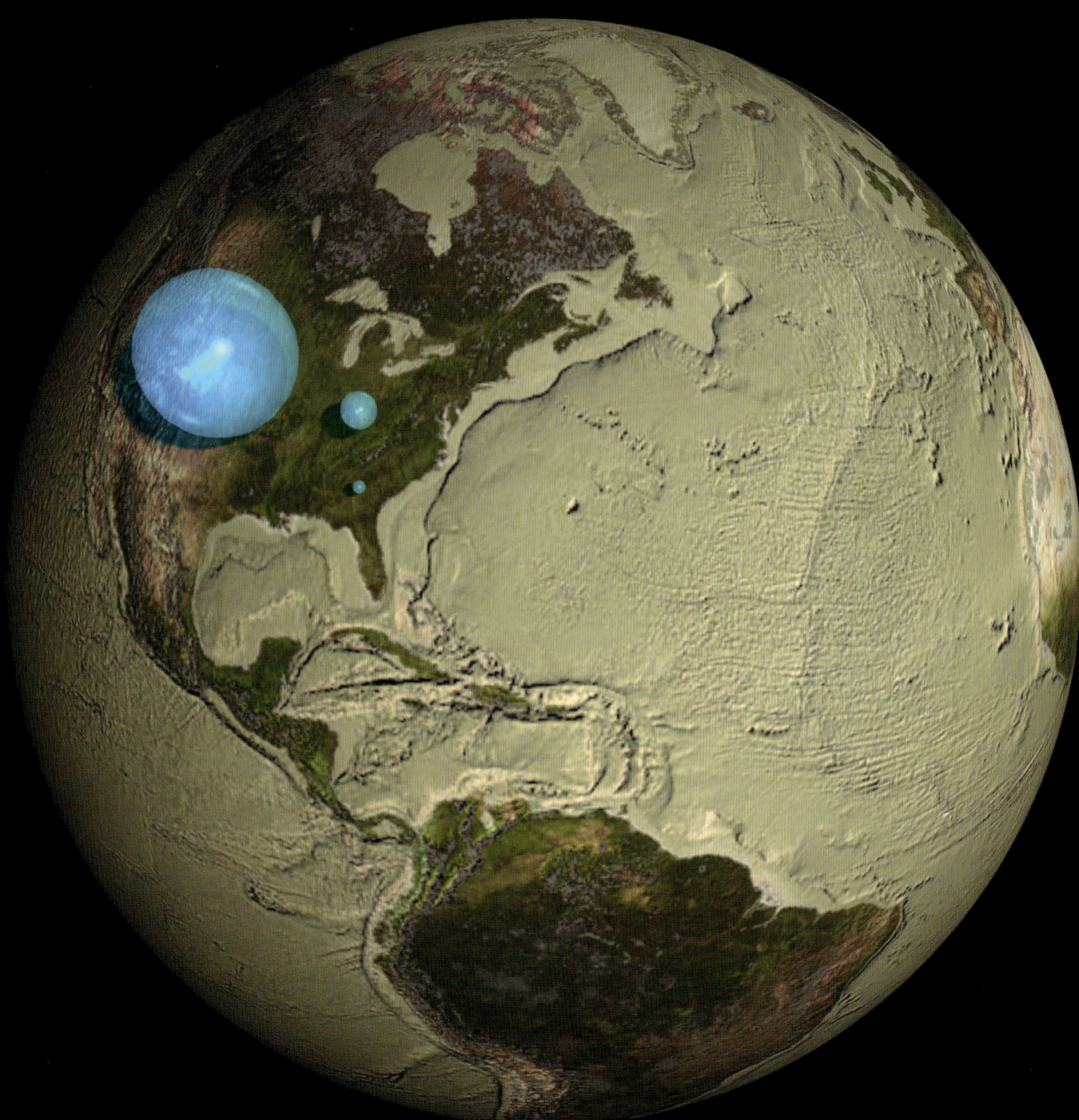

All the water in the ocean, the atmosphere, groundwater and surface water could all be rolled into a ball just 1,500 km across. Only about 3% of the world's water is fresh (the mid-sized sphere) and of that only one-third is easily accessible to humans (the smallest sphere).

Big ocean – little water

Prepared at the Woods Hole Oceanographic Institution in Massachusetts, the image on the facing page shows an accumulation of all the water in the ocean, rivers and seas, in groundwater and in the atmosphere, combined into an imaginary ball. It would measure just 1,500 km across.

I now realise that when we dive in what appears to be an endless ocean, we are diving within that little blue ball – it hardly seems endless now. That is a stark reality – the facts versus the mythology of *The Blue Planet*. This one image certainly makes the ocean seem even more precious.

The stark reality of the eternal ocean, just a thin film covering the earth, more precious than one can imagine.

All creatures in the ocean exist within that little blue ball, all in dynamic balance. Even when nature tips the balance with the eruption of a local volcano or an unusually severe El Nino effect, the scale and dynamism of the overall oceanic ecosystem prevents global catastrophe. This astonishing ability of complex systems to regulate themselves is called *homeostasis*.

Complex systems and unpredictable consequences

The waters of the ocean are never still. Surface currents are heated by the sun's energy at the equator and then run around the continents, influencing the weather and with it the adjacent land, agriculture and human settlements. Water flowing towards the poles cools and sinks, travels slowly along the sea bed and into the darkest abyss, until gradual warming causes it to rise again. These vast *conveyor currents* mix and blend all the ocean's waters in huge continuous loops.

The ocean – a complex system kept in balance by homeostasis.

At every level, down to the smallest microworlds that surround sea anemones, parts of the system are acting and interacting constantly, creating layer upon layer of life so complex that they cannot be fully understood, or predicted. Just think of the ocean's effect on the weather – small changes can have consequences that overwhelm even the world's most powerful computers as they try to predict next week's weather.

The ocean is the key driver of the world's weather. When it changes, climate changes.

Vertebrates apart, the vast majority of life in the ocean and much of it on land is doing just as well now as it was millions of years before we showed up. When viewed from a long-term perspective, the majority of life on the sea floor has not just *survived* from *ancient times* but has been *doing just fine*. Corals, for example, seem just as happy in the last ten million years as they were 200 million years ago – or at least they were until we started burning carbon faster than the system can handle it.

Climate change – local vs global effects

The ocean is not just the womb of all life – it continues to be its engine, creating and controlling the physical cocoon in which all life exists.

Homo sapiens was named in 1758 by Carolus Linnaeus, the father of modern taxonomy, and is the only species in the *Homo* genus that is not (yet) extinct. *H. sapiens* has been around for about 315,000 years and this subspecies of anatomically modern humans, *Homo sapiens sapiens*, is thought to have been around for 35,000 years – and haven't we done badly! Just sample a few of typical headlines from the last year:

> "*Homo sapiens* has now come to replace Nature as the dominant environmental force on earth."
> Lewis Dartnell in *Origins: How the earth made us* 2018.

> "Terrifying climate change warning: 12 years until we're doomed."
> New York Post, 8 October 2018

> "The collapse of our civilisations and the extinction of much of the natural world is on the horizon."
> David Attenborough at Davos, February 2019

When we refer to *Climate Change* it usually implies *Anthropogenic Climate Change* – the influence that mankind has had in just the last few hundred years.

Five hundred million years of climate change

Climate change, often at dramatic levels, has been a part of all history and prehistory, as this chart shows...

This chart shows that the long-term evolution of oxygen isotope ratios as measured in fossils. They reflect both the local temperatures and global changes associated with continental glaciation. As such, these ratios can be interpreted as rough changes in climate.

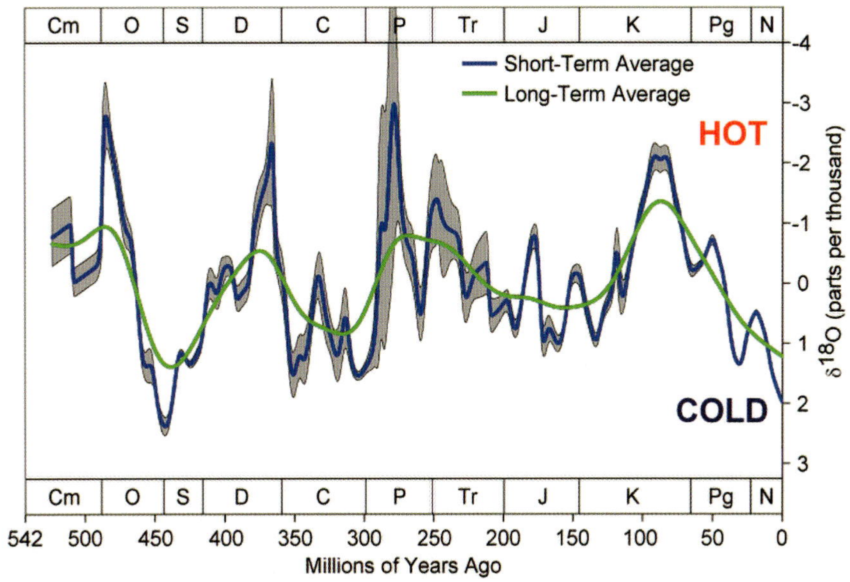

Image from the Global Warming Art project. Based on Veizer et al. (1999 and 2004).

There is nothing 'perfect' about our current climate – it's just what we've become used to in our short history on earth.

It must be said that these results are only a rough indication, and some suggest that the highs and lows in the early half of this 500 million year period were both significantly warmer than shown here – earth was an almost unrecognizable greenhouse planet. After the sweltering climate that launched the age of the mammals, we currently live in a relatively 'cold' phase.

Some billion years ago it rained incessantly for more than a million years. One hundred million years ago Europe was a tropical swamp unsuited for human habitation. Just twenty thousand years ago most of northern America and northern Europe was covered in an ice sheet several kilometers thick. We should not forget that we live in one of the most benign and stable periods in the planet's tumultuous climate history.

The entire range of living matter on Earth, from whales to viruses and from oaks to algae, could be regarded as constituting a single living entity capable of maintaining the Earth's atmosphere to suit its overall needs and endowed with faculties and powers far beyond those of its constituent parts.

James Lovelock
Gaia: A New Look at Life on Earth, 1979.

To fundamentally disrupt weather systems would take a massive and sustained effort spanning centuries. Unfortunately, that is exactly what we are doing.

To fundamentally disrupt these complex mechanisms and thereby change conditions in the ocean permanently would take a massive and sustained effort spanning centuries. Unfortunately, that is exactly what we are doing.

Over hundreds of millions of years, massive reserves of plant-derived carbon were deposited underground as coal – the shallow seas of the Carboniferous Period were a great example of ideal conditions for this – while the bodies of oceanic microorganisms have been liquefied into petroleum and gas. In our modern world, we humans have been using up this concentrated fossil energy – made up of countless trillions of living things – to improve our lives.

Colossal amounts of carbon dioxide released into the air through the burning of these fossil fuels will have profound effects on living things in the ocean. The water temperature is raised as the Greenhouse Effect warms the atmosphere, and the ocean's natural alkalinity is reduced as carbon dioxide dissolves into the water to form carbonic acid. This acidification may get way beyond what corals and other marine life can tolerate, with serious consequences for the ocean.

While media images of human-caused effects on the ecosystem are dramatic – tons of plastic accumulated in mid-ocean whirlpools, factories pouring chemicals into rivers, dead seabirds on oil-blackened shores and treeless hillsides loading rivers with topsoil runoff – they are essentially examples of *local* effects.

The *global* consequences of releasing concentrated fossil carbon, while less immediately visible, are on a far larger and more dangerous scale, as their effects could be *systemic*. At some point, natural homeostasis is unlikely to be able to handle it and a small proverbial straw could break the camel's back. In modern scientific parlance, it's all fine until it suddenly isn't – until the complex system experiences a *bifurcation point*; until something happens that causes an effect that resonates throughout the system, such as extreme climate change, quick and profound enough to result in a large extinction.

It's only 200 years since the industrial economy began. Most of the damage we have done to the environment has been done during that time.

We will reach a point where natural homeostasis is unlikely to be able to handle all the impacts of climate change.

For the last few centuries we saw nature as the 'means of production' – resources such as water and coal that we could utilise for our own ends. Our industrial world is only 200 years old and most of the

damage we have done to the environment has been done during that time. Agriculture also has a lot to do with it. Almost half of the earth's land has been converted to farmland and more than half of mammal biomass is now made up of farmed animals. Studies abound on the impact of commercial fishing, claiming that up to 90% of large ocean animals have been removed since 1950 and that we will run out of wild-caught seafood within 30 years.

The only hope we have for the future is that we are not yet beyond the tipping point – that we're not yet in a mass extinction event.

It's certain that we're pushing nature's complex systems close to what's sustainable for our own good. What we have learnt from Chaos Theory about the behaviour of complex systems may come back to bite us here. Academics talk about impressive concepts such as butterfly effects, emergent properties, bifurcation points and Black Swans. Everything's fine until suddenly it's not and the whole system breaks down.

With new ways to measure climate change our understanding is constantly evolving.

The Sixth Extinction?

Today, some say that the evidence indicates a sixth mass extinction is under way. They call this the *Anthropocene Extinction* and, as the name suggests, lay the blame for this one firmly on the shoulders of humans. Some projections say that by the year 2100, human activities such as industrial pollution, land clearing for farming, and overfishing may drive more than half of the world's marine and land species to extinction.

"Like all extinction events, so far, this one has been phased and complex, spanning tens of thousands of years and starting when our kind left Africa", says Smithsonian palaeontologist Doug Erwin. However, he thinks talk of a *Sixth Extinction* is "junk science".

"Many of those making facile comparisons between the current situation and past mass extinctions don't have a clue about how truly awful the mass extinctions recorded in the fossil record actually were." He is not trying to downplay the destruction wrought by humans, he just believes scientists like him have a responsibility to be accurate about such comparisons. "If we're really in a mass extinction then there's no

Less than one percent has gone extinct in recent human history.

point in conservation biology; it's not a matter of saving tigers and elephants – go get a case of scotch," he said.

"How many geographically widespread, abundant, marine taxa have gone extinct thus far? The answer is, pretty close to zero," Erwin points out. In fact, somewhere less than one percent of the stony corals, amphibians, birds and mammals have gone extinct in recent human history.

Saving the planet

In recent years it has become popular and politically correct for well-to-do individuals and businesses to care about potential extinctions faced by some of the more 'romantic' life forms on our planet such as the rhino, elephant, tiger and the panda. Together these are sometimes termed the *warm and furries*. They are certainly iconic and typically get more attention than the *wet and slimies* which always struggle to be taken seriously. We don't really care about them until their possible absence might affect us directly.

Idealism increases proportionally with ones distance from the problem.

John Galsworthy

In the 1970/80s the stand-up comic philosopher George Carlin used to rail on about how arrogant and self-important we humans were in our attempts to *Save the Planet*. I think he got most of that right and certainly put us into our place.

"Save the planet! Are these people kidding? We don't even know how to take care of ourselves – or for one another. The planet has been through a lot worse than us. Been through earthquakes, volcanoes, plate tectonics, continental drift, solar flares, sun spots, magnetic storms, the magnetic reversal of the poles, millions of years of bombardment by comets and asteroids and meteors, worldwide floods, tidal waves, fires, erosion, cosmic rays, recurring ice ages … and we think some plastic bags and some aluminium cans are going to make a difference? By the way, there's nothing wrong with the planet in the first place.

The planet is fine. The people are f….d!"

It is ironic that during that time, when some students used his conclusions in their work, they were kicked out of the classroom for their radical views.

Today, we buy disinfectant products that promise to "Kill 99% of bacteria". Really? Remember that for every cell in your body there are more than ten bacterial cells. Without the bacteria and microbes integral to the functioning of your body, you would be stone dead anyway.

If we're honest about it, selfish arrogance is what drives our concern for the planet and the ocean. Our primary instinct is to make sure that the planet will be better for US and OUR kids – not even for ALL children.

We don't need to save the planet, it's looked after itself for four billion years. It's always been habitable, and things have lived on it, so why worry?

James Lovelock,
Dorset Daily Echo 23 July 2016

What are we really trying to achieve?

Mostly, in our endeavour to *save the planet* we are essentially trying to keep things the same as they are, or to bring back the past – the way things used to be. Could the word 'conservation' be getting in the way of innovation, development and opportunities for new life?

Does conservation, as an attitude to life around us, prevent opportunities for new life to evolve?

Our business world seems to have a great appetite for *innovation and growth*, while as individuals we seem hell bent on keeping things the same, conserving the status quo, preventing change in the environment, nature and the weather. We strive to keep things as they are and do even that with little common purpose or success.

Change threatens the status quo.

Extinction: Creative destruction

In my business career I advised corporations on how to avoid commercial extinction in the face of disruptive new technologies. This is roughly the equivalent of consulting with dinosaurs (the top predators) on how to thrive in the face of a radically different future environment. As one would expect, I stood on the shoulders of giants who thought about this challenge long before me.

In 1922, Russian economist Nikolai Dmitrijewitsch Kondratieff published an essay on *Long Economic Cycles*. This study of the long cycles in capitalist economies (*long* in this case was deemed to be 40–50 years) did not find favour with his superiors; and Kondratieff ended his life in Siberian exile.

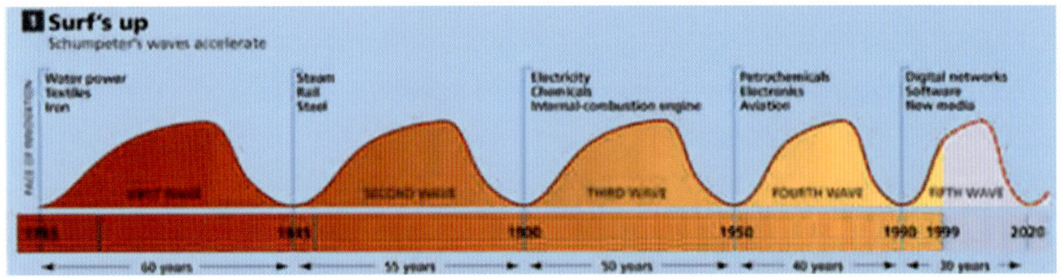

The Economist, February 1999.

I've been consulting with corporate dinosaurs (the top predators) on how to cope with radically different future technologies (a change in their ecosystem).

Yosef Schumpeter, Harvard professor and one-time Austrian Minister of Finance, believed the driving force behind the observed *Kondratieff Waves* was innovation. In his book *Business Cycles*, published in 1939, Schumpeter associated each of Kondratieff's Long Waves with specific innovations in technology and commerce – steam power, petrochemicals, digital technology etc. Every major wave of innovation brings with it new economic activity and often new growth. The more profound the success and the steeper the growth, the greater the risk of accompanying destruction – of old competitors, of cherished old ways of doing business and of old ways of working, and jobs.

Schumpeter envisioned economies whose growth was propelled by the entrepreneurial spirit and breakthrough technologies. He coined the term *Creative Destruction* to describe the disruptive effects of innovation. In my book on *Radical Innovation* I concluded that "any business leader who seriously wants to lead through innovation has to be ready to cope with and manage the rather more destructive consequences of innovation".

Putting extinction into context

Today I find the same process could apply to our thinking about waves of probable, possible and imagined 'extinctions'. As we have seen with the *Great Oxygenation Event* earlier in this book, over a period of two billion years cyanobacteria poisoned existing anaerobic life on our planet by flooding the atmosphere with oxygen. In fact, what has been called *The Biggest Environmental Disaster of All Time*

The business world seems to have a great appetite for innovation and growth. As individuals and nations, we seem to be hell bent on conserving what we have – to avoid change.

simultaneously made the planet habitable for us, and all other aerobic life. Creative destruction QED.

There are two important evolutionary events that ensured the emergence of humans: the rise of fish and the extinction of the dinosaurs.

If earth had been a democracy 3.5 billion years ago then all anaerobic beings would have voted for new environmental laws to prevent the accelerating levels of oxygen in the atmosphere. At the time of the extinction of the giant dinosaurs the largest mammal was the size of a shrew. Mammals evolved into the white space left by the extinction of these giant reptiles. We humans would not be around today were it not for this particular extinction. Thankfully, and I say this with an unbridled anthropocentric attitude, the dinosaurs never were given a vote to prevent their extinction. And, I'm sure that the Nautilus sorority would have lobbied against the acidification of the oceans during the end-Cretaceous extinction.

If the past 500 million years has taught us anything it should be that extinction is a vital and an essential part of evolution.

If you step back from the 'problem' of extinction, it is clear that everything around us today, including ourselves, is the direct result of several major past extinctions. The only reason to fear extinction is if you take it personally. This is what Geoffrey West calls *The Curse of Consciousness*. We all know we are going to die, and that's our burden. Death is a natural and important part of life – there would be no new development without it – and so is extinction.

Just as death is a part of the cycle of life and as surely as spring follows winter....so extinction creates the opportunities for the development for new and different forms of life.

Life today is a consequence of all previous extinctions.

All progress, including learning itself, requires constant *letting go*, It's about death, and re-birth. It's about learning and internalising something new. Change is constant and accelerating. Sometimes it feels as if we have to change as fast as we can just to maintain our current status quo, just to stay in the same place.

Extinction is a vital and essential part of evolution.

If we were 'gods', with a proper 'global' or 'universal' perspective, we would likely welcome extinctions as just another example of *Creative Destruction*. Perhaps a better way to think about *extinction* would be *Creative Extinction*.

Is extinction ok?

In November 2011, *Conservation Biology* published an article that reports the findings of a survey that asked if it is time to let some species become extinct? Sixty percent of conservationists agreed that specific criteria are needed to determine whether a species is worth saving. As shocking as it may seem, most conservation charities prioritise efforts on the basis of a checklist system already.

Are all species worth saving?

Should this mean ultimately withdrawing support for certain species? If so, at which point do we draw the line? And within what context do we make such decisions? Who will make those decisions? And, who will give them that authority?

We cannot destroy life on earth, nor stop its inexorable evolution. If we choose to worry it should be first and foremost for our human family, for it is we who are most at risk.

Robert Hazen *The Story of Earth* 2012

Certain extinction?

One thing is certain – there will be another major extinction event – the only question is when, and who will survive. Will we? Should we be concerned? Are we? Perhaps we should be, but what chance do we have of survival if we are more concerned about the possible demise of the African elephant and calling for the destruction of 200-year-old carved ivory heirlooms, than our own fragile position on this planet.

In 2003 Sir Martin Rees, published his book *Our Final Century* – his scenarios for the many unintended consequences of our advanced civilisation, and why it was unlikely that the human race could survive the next century. It arrived to much critical acclaim but little popular attention. That year the best sellers were the *Harry Potter* books, *Dr Atkins Diet* and Bill Bryson's *A Short History of Everything*.

Biology knocks humans off a pedestal. We're not the pinnacle of evolution, but part of nature.

Alice Roberts, *Sunday Times Magazine*, December 2018

The end of the human race will be that it will eventually die of civilisation.

Ralph Waldo Emerson

Clearly the really big issues struggle to get into our attention span. The frog in the slowly-heating pot of water comes to mind. Elon Musk says he wants to establish a colony on Mars, and that he plans to die there. By implication it means that mankind's ambition (or perhaps Musk's) to leave this planet is stronger than the commitments to survive here. Perhaps if we humans can't work out how to live sustainably in this existing and highly favourable environment, then we have no right to try to change the answer by changing the question.

If this book has shown you anything in this regard, then it must be how modest a part of life on this planet we actually are and how miniscule the time is that we have spent here. And, how arrogant we are about our importance.

This madness is in the capacity of the human mind's urge to do more than is strictly necessary for its survival.

Bryan Appleyard commenting on Elon Musk, Sunday Times, 11 February 2018

An evolutionary perspective of our place in the history of the earth reminds us that Homo sapiens sapiens *has occupied the planet for the tiniest fraction of that planet's four and a half thousand million years of existence. In many ways we are a biological accident, the product of countless propitious circumstances. As we peer back through the fossil record, through layer upon layer of long-extinct species, many of which thrived far longer than the human species is ever likely to do, we are reminded of our mortality as a species. There is no law that declares the human animal to be different, as seen in this broad biological perspective, from any other animal. There is no law that declares the human species to be immortal.*

Richard Leakey, Paleoanthropologist

Immortality?

The ability to live forever may seem like a desirable option, at least until you begin to think seriously about the consequences.

Some animals don't seem to age at all – or do so very slowly. *Ming the mollusc* is the oldest verified solitary animal on record. This ocean bivalve was 507 years old when biologists dredged it up from the coastal waters around Iceland in 2006 and it was dated by the growth rings in its shell. Of course, *Ming* did not survive the effects of this rigorous scientific study, but one assumes there are many more life forms like it down there.

When Turritopsis *is stressed it can revert to the polyp stage on demand, being* reborn *as a young clone of itself.*

Colonial animals like corals can survive for many thousands of years. However, individual coral polyps may be only a few years old. Others, like some jellyfish, don't reproduce sexually, and instead are able to make tiny clones of themselves – they are called *biologically immortal*. Of course, these organisms are definitely still mortal. They can be killed by a predator, disease, or an environmental catastrophe.

An adult *Turritopsis*.

One extreme example of this is *Turritopsis*, a tiny hydrozoan, aptly nicknamed *The Immortal Jellyfish*. It begins life as a free-swimming larva, then settles down and gives rise to a colony of polyps on the ocean floor which bud off and live as free-swimming jellyfish. The polyps and the jellyfish arising from a single colony are genetically identical. When *Turritopsis* is under stress, is sick or old, it can revert to the polyp stage on demand, forming a new polyp colony and essentially being *reborn* as a young jellyfish. Theoretically, this process can go on indefinitely, effectively rendering the jellyfish *biologically immortal*, although in practice most are likely to be eaten without reverting to the polyp form.

This capability makes *Turritopsis* an important subject for biological and pharmaceutical research, such as the quest to find a way to make stem cells for renewing damaged or dead tissue in humans.

A juvenile *Turritopsis*.

'These tests are the most violent thing we've ever done to the ocean. It's equivalent to 216 Empire State Buildings being blown into the sky.'

What if we did our worst?

Between 1946 and 1958 the United State detonated a series of 23 nuclear devices around the Bikini Atoll in the South Pacific. One of the largest explosions vaporised three islands and created a massive radioactive crater. The explosion shook islands hundreds of miles away, generated a wave of heat measuring 99,000°F and spread radioactive fallout as far as Japan and Australia. This bomb was 1,100 times more powerful than the atom bomb dropped on Hiroshima. Quite dramatic.

Now, 60 years since the test ended, it appears that ocean life has dramatically, and somewhat surprisingly, recovered.

"Stanford scientists are amazed by thriving marine life at Bikini Atoll site where 23 atomic bombs were dropped" screamed the headlines after a 2017 study findings. The atoll has an abundant ecosystem of plant and animal life, corals as big as cars, teeming with snapper, sharks and tuna. Also, crabs were said to be feasting on coconuts filled with a radioactive groundwater. The abundant coral seemed to be immune to extremely high radiation levels. Steve Palumbi, director of Stanford's Hopkins Marine Station, declared life to be "remarkably resilient" in adapting to radiation. Life adapted, evolved and recovered, even after these dastardly experiments.

After some 60 years, Stanford scientists are amazed by the thriving marine life at Bikini Atoll.

Evolution is not just opportunistic, but in the face of the biggest onslaught anyone could have thrown at it, ocean life has demonstrated unimaginable resilience and surprising continuity.

It appears that even if we chose to detonate all the nuclear devices and nerve agents we have, it will not eradicate life on earth. Perhaps the only species such an event would kill off is our own. It would certainly be a mass extinction from our perspective. But, if the End-Cretaceous or End-Permian events did not stop evolution, I doubt whether anything we could do would either.

One thing we need to be careful about is attributing supernatural agency to ourselves, the human species – that's what got us into the current mess.

What if we did our best?

If we did try to do serious good, what could we do? Could we step out of our conservative ideals, think differently and do something radical? Fortunately there are some great examples of radical innovation – here's one.

Coral reefs support about 25% of ocean life while covering less than 1% of the seafloor. And, as we all know they are threatened, nowhere more so than around the Florida Keys. Last summer I was attending a conference in Key West and was invited to view a coral restoration project being run at the Mote Marine Laboratory. Here David Vaughan and his team came to a counter-intuitive conclusion – coral thrives

Previously it took six years to produce 600 corals in the laboratory – now it can be done in an afternoon.

Improving on nature, in their lab corals reproduce up to forty times faster than in the wild. They're teaching reefs how to remain eternal.

better when it is damaged. In the wild, hard corals take many decades to reach sexual maturity – up to 75 years for some species. Using their microfragmenting approach they grow corals in trays and then break the juvenile corals into small clumps of just a few polyps. This reduces time to sexual maturity to just three years – the effect is that corals reproduce up to forty times faster in their lab – after which they are planted back onto the damaged reef allowing it to recover – and to remain *eternal*.

This breakthrough discovery prompted Vaughan to postpone his retirement. "Previously it took us six years to produce 600 corals in the laboratory – now we can do that in an afternoon. Our aim is to put millions of corals out there in the next few years," says Vaughan.

This discovery was another of those happy accidents that happen when passionate individuals keep their eyes open to the *unlikely* – even to things previously thought *impossible*. It's a different way to think about the future of the ocean. This is radical innovation in action.

David Vaughan, manager of the Coral Restoration program at the Mote Marine Laboratory.

We only have to look at ourselves to see how intelligent life might develop into something we wouldn't want to meet.

Stephen Hawking

Two big questions

Towards the climax in Dan Brown's book *Origin*, the hero Robert Langdon's friend Edmond Kirsch declares that two fundamental questions have captivated humans since we first became self-aware:

Where do we come from?
Where are we going?

These are questions about our *origin* and our *destiny*. In terms of the former, I hope that this book has covered that subject, at least for the past few billion years. Before that, it's a little uncertain and all we can lean on is a little quantum physics and the Big Bang.

We really have no idea where we are going. It would be nice if we had a common will – a shared strategic view, a picture in our minds of a desired outcome for our place on this planet.

Nations can't seem to agree on common agendas for the most important issues, be it the ozone layer, climate change or plastic pollution. Denial seems to be the default option for these proverbial *elephants in the room* that demand global consensus. Just take the case of countries such as the USA, Saudi Arabia and Russia refusing to be a part of the Intergovernmental Panel on Climate Change declaration in October 2018. Whatever happened to the great commitment to global cooperation and open borders that would catapult the whole world into a universally prosperous future?

On top of what we already know, there are new realities being postulated around the concept of *Climate Departure* – that a 4% rise in overall global temperature would make large chunks of the world uninhabitable within 20 to 30 years. The consequences could spark a flood of human migration from central America, Africa, central Europe and Asia to cooler and richer parts of the planet – making the current global migration issues seem quite minor. If we can't agree on a limit of a few degrees of global warming, how will we then agree on an approach to finding a better home for perhaps a billion displaced people?

It's always about us

How do we see the world and all life in it? Undoubtedly from a human-centric, anthropocentric perspective. In the end we will always put ourselves first.

Our questions concern how things will affect *us*. How will climate change affect me? What damage are we causing? Is it worth it? What animals can we save to make us feel better? We're not really interested in *The Planet*. We act on the idea that someday in the future we might be *personally* inconvenienced. The uncomfortable reality is that our decisions about the future will always be anthropocentric – it's only natural. All life on earth puts itself first.

Yes, we humans are influencing the planet – but sometimes, in our arrogance, we tend to forget that we are an integral part of *the system*. Aren't we as *natural* as any other creatures on this planet? We are as much a part of evolution as anything else. Each and every one of 'them', the other animals, has an effect on the environment in which they live (or have lived) and we never complain about the effect 'they' are having on the environment – unless it affects one of 'us' directly.

Unfortunately our intelligence may be contributing to the problem. Paul Callomon, who was an integral partner in the conception of this book, articulated this deeply inconvenient truth: "The cause of many extinctions was a runaway phenomenon of some kind – resulting in climate change etc. Its effects became exponential and by the time it had burned itself out and homeostasis returned the planet to a stable condition, lots of things had died out. Human intelligence (not knowledge or wisdom, which are different) is just such a phenomenon. It grew gradually but then ballooned with the advent of writing, which allowed humans to tell each other in advance how to be intelligent. In terms of the global ecosphere, therefore, our intelligence is a disaster on a par with some of the previous ones".

Selfishness & altruism

Published in 1976, Richard Dawkins' book *The Selfish Gene* went on to be rated as *the most influential science book of all time* in a 2017 Royal Society poll. Dawkins wanted "to examine the biology of selfishness and altruism". He asserted that "gene selfishness will usually give rise to selfishness in individual behaviour. However, there are circumstances in which a gene can achieve its own selfish goals best by fostering a limited form of altruism at the level of individual animals."

Sometimes one can achieve one's own selfish goals best by fostering a limited form of altruism.

We humans frequently adopt the same strategy to expedite what we want to achieve. It is also the essence of *Ubuntu* – a Nguni African term meaning *humanity*. It is often translated as "I am because we are" and articulates a cultural philosophy and world view that has become synonymous with the best of African attitudes.

Having previously spent a few wonderful decades in that part of the world, I offer here a 21st century interpretation of Ubuntu based on my personal experience living this world view. I believe it could be a guiding principle for our relationships with our fellow humans and other life on earth.

My Ubuntu.
We don't have an independent existence
but are an integral part of the context within which we exist.
It's our connectedness that creates meaning.
We are powerful independent individuals,
yet when we look for the individual within us,
we find it only in the networks of our relationships.
To gain our full humanity,
we have to restore our connectedness with the entire web of life.
The success of any community depends on
the personal success of the individual members.
The ultimate success of the individual
depends on the success of the community as a whole.
We are ubiquitously connected individuals in a networked society.
Each of us is the one we all depend on.

Thinking about the future

I have always found it strange that relatively few palaeontologists and fossil collectors are interested in the living equivalents of the fossils they study, or that few scuba divers or marine biologists care much for the fossil ancestors. It was even suggested to me by one biologist that "life begins where fossils end".

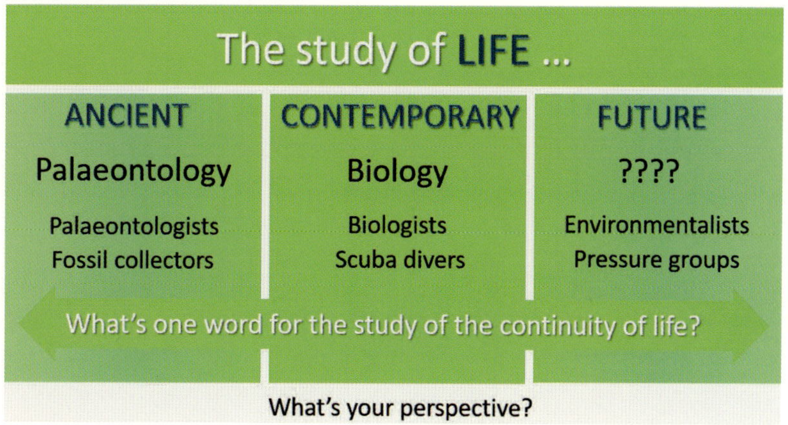

As Paul Callomon pointed out in the Preface to this book, we haven't yet found that elusive word that can encompass the study of all life, covering the ancient, the contemporary and the future. Even so, there is an important point that remains here. By no longer drawing a distinction between the ancient, contemporary and future life, we can appreciate the true continuity of life and properly locate humanity within life's vast story – in which we form the latest and so far, fleetingly brief chapter.

Only when we fully appreciate the sheer scale of life on this planet, and our place within it, will we conduct ourselves responsibly towards life on earth, and in the ocean – the womb of all life. While restraint may be an alien concept for humanity, we will have to think critically about the future, not out of altruism but from enlightened self-interest.

When you fully appreciate the sheer continuity of life on earth and humanity's part in it, will that change your attitude to the future?

The ultimate uncertainty comes from someone asking a really new question.

The future is not inevitable, nor is it one ordained destiny. Out there are many possible futures.

Thankfully things are not certain. We will keep on living in an uncertain world. Think of each uncertainty as a moment of perfect freedom. A moment full of possibilities. Imagine the alternative, a life stretching before you where everything is predictable and certain. That's not life at all.

Homo sapiens is unique in being able to consider alternate and potential future scenarios. No matter how difficult the options, we always have a choice. We can consider and learn from new facts, and change. We can act appropriately. We can evolve.

Our future is a matter of choice, not chance.
The primal spirit of life will go on.
The ocean will thrive.
But we are just visiting this planet.
Tomorrow belongs
to those who can
hear it coming.
David Bowie
We shall not cease from exploration
And the end of all our exploring
Will be to arrive where we started
And know the place for the first time.
T S Eliot
It is not in the stars to
hold our destiny
but in ourselves.
Shakespeare

FOUR SURPRISES

What's been learnt?

We are insignificant in the biomass of *LIFE*

- Plants are the biggest component and represent more than half, but are mostly absent in the ocean
- Bacteria account for almost all of the other half, were there before us and will survive us all
- Animals represent less than 0.1% of life's biomass

We have little *TIME*

- On earth's 24-hour clock we've been here less than one second
- We are amongst the most fragile of all life forms
- Having been 'spoilt' by our technology, we can hardly survive without it
- This may be our last millennium

Our blue planet has a very small *OCEAN*

- It's the womb of all life
- It's small and finite
- All life on earth will be shaped by its future health

CONTINUITY trumps extinction

- 90% of major life forms have existed more than 300 million years
- Without extinction there is no new white space
- The ultimate top-down view would see life on earth as a very successful experiment

Nautilus
Heteromorph
WOLFGANG
MMONOIDEA
OF THE
LOWER
GREENSAND
ATLAS
TEXT
Aldafira
WOLFGANG GRULKE
MAY 2017
INTO LAND
INTO WATER COLUMN
CONCEPTUAL PHYLOGENY OF TODAYS

Wellesley Duke of Wellington
1814 K.G. – G.C.B.
TRIASSIC
JURASSIC
TODAY'S SURVIVORS

School

THANKS

I am deeply indebted to many friends, new and old, who have helped in so many ways to bring this personal journey to life. They have been my mentors one and all, some in really surprising ways.

I want to thank a few in particular: Paul Callomon for his inspirational and critical debate when this book was being conceived, and his laser-sharp attention to detail when we were working on the final structure and content. Tom Kapitany, always generous with his time, passion, insight and hands-on experience – often delivered from the most obscure parts of our planet. Alberto Rava, Anton Musgrave, Stef and Fiona Kriel with whom I have shared so many experiences, from alternative futures to marine expeditions in the world's most remote oceans. Their images help illustrate my story.

John Whicher, Bob Chandler and Phil James continue to inspire me with their enthusiasm for Jurassic fossils. Dean Lomax generously shared his deep passion for, and knowledge of, the Sea Dragons. Three artists helped illustrate the unimaginable - Richard Bizley, Karen Gossel and Ray Troll. For bringing my random memories into cohesive design, Jenny Penny and Prof. Phil Cleaver, you did us proud. My thanks also to the many friends around the world who have shared their ideas and photos in person and on social media (their contributions are individually listed elewhere), and those who have taken the personal time to review and criticise – their help and encouragement was truly invaluable. Finally to my wife Terri, who makes everything delightfully purposeful for me.

At this point in the book we could have included images of the most important contributors, as we did for the other books in this trilogy. Brilliantly our design team at et al suggested that we invite our contributors to send us pictures of their 'kids' of all ages – literally the next generation to enjoy this eternal ocean – and to whom we will bestow this fantastical world.

And here they are… they will have more time than we do to enjoy the present and to make better choices for their future.

SUBSCRIBERS TO THE FIRST EDITION

These are the Angel Investors whose confidence in the final product far exceeded any physical evidence. Their faith and encouragement spurred me on every step of the way.

Albert G. Sheean
Alberto Rava
Alexander Schmalix
Amy Tripp
Andreas Grünemeyer
Andreas Schmidt
Andy Steadman
Anthony Lindgren
Anton Musgrave
António Monteiro
Bronwen & Karl-Heinz Töpfer
Bruce Carlson
Burger Collection
Charles E. Rawlings
Chris Tait
Christian Meyer
Craig Longbottom
Darren J. Brown
David Baldwin
David Beardsmore
David Groocock
David Harrison : HRH Geology
David Velinsky
Diana Natalie Conway-French
Dmitry Vinogradov
Dr. Jochen Gerber
Dr. Herter László
Eric & Isabelle Lancrenon
Felix Lorenz
Frédéric Meunier
Gale A. Bishop
Geoffrey Cornwel-Smith
Gregory Jeff Barord – Save the Nautilus
Gudrun Humpohl & Ralf Metzdorf
Guido Poppe
Ilse Braun
James and Georgie
Janet Taylor
Jay Armstrong, Elementum Journal
John & Jenny Whicher
John & Moira Lauderdale
John Gaye
John McLoughlin
Jonathan & Lucy Roberts
Joseph & Cassandra Huston
Joseph Sullivan
Julia & Brian Whelan
Julian Joseph
Karen Gossel
Kelsey James
Ken & Karenza Mathews
Kenneth Ho
Leslie K. Burr
Luis Fernando Espinel Martinez
Lynda & Simon Armstrong
Malandrinos Family
Marten L. van Straten
Martin & Sarah Snow
Murray Edmunds
Nick Hawes
Oleg Kuznetsov
Oliver & Harry Press
Oliver Frank
Patrick Branger
Paul Callomon
Paul Le Marquand
Paul W. Fletcher
Peter Coldicott
Peter Thomas
Philip Hadland
Philip R James, Southampton
Philippe Poppe
Philippe Vandenberghe
Pierre Ferruit
Prof. Christopher Bartlett
R M England
Ramsés Moreno
Ray & Angie Culver
Richard Bizley
Richard Dorrell
Rick Rolater
Rick Ross
Sam Caethoven
Sam Simons
Skyler Phelps & Lynn Fischbeck
Steve Etches
Sue & Nigel Masters
Thomas Billert
Tim Haye
Tom Kapitany
Trefor & Sheila Harries
Trefor Harries
Trevor & Christine Winer
Vincent J Valle
William & Helen Rae
William Blakers
Xavier Simona Sucarrats
Шиловский Олег

IMAGE CREDITS

Photographs, drawings and charts are primarily by Wolfgang Grulke, plus these contributors. Page numbers are indicated with B=bottom, T= top, M = middle, L= left, R=right, C=centre, W=Wikipedia/Wikimedia

Alberto Fernandez Fernandez/W: 12/13
Alberto Rava: 23/24 u/w photos
Alexander Semyonov, MSU/W: 152R
Andrea Marshall, QueenOfMantas.com: Front cover, 127BL
Anthony Lindgren, www.LindgrenFossils.com: 112BL, 183TL
Anton Musgrave: 16/17, 131BR, 133TR, 145TR, 146TR, 161B, 173CR, Back cover
Bernard Picton/W: 165BR
Charles Messing: 120
Charles Rawlings: 157TL, Collectors' Edition slipcase image
Christian Klug: 98TL
Christopher R. Scotese, www.scotese.com: 22
DeDuijn/W: 34BR
ESA/W: 47
Eye of Science: 167
Francois Escuillié, Eldonia: 95, 116TL, 168C, 174TL, 178TC
Geoff Spiby: 37TL
Hans Fricke & Schauer/W: 172BL
Jacek Major, PaleoArt: 98BL, 190/191
Karen Carr: 109BL
Karen Gossel: 78 Colour design
Kevin Guertin/W: 187T
Luc Ebbo: 108
M. Buschmann/W: 165TL
Marianne Cornelissen-Kuyt/W: 151BL
Marie-Lan Nguyen/W: 54TR
Mark V. Erdmann/W: 173TL
Mark Tatchell: 216
Mark Walz/W: 149BL
Michelle Pemberton/W: 138BR
Mike Reich/W: 150BR
NASA/W: 23R, 194, 195
NPG London: 68R
NOAA, NOAA Emma Hickerson/W: 35R, 124TL, 144TR
Nobu Tamura/W: 169BL
Paul Callomon, The Academy of Natural Science, Philadelphia: 18, 20/21, 68BL, 151BR, 154BL, 156TL & CL
Paulyn Cartwright et al./W: 125
Peter Ward: 118BR
Phil James: 135BR, 136L
Ray Troll www.TrollArt.com: 60, 73, 114
Richard Bizley, Bizley Art: 81L, 82, 142BC, 198BL
Richard Dorrell: 57BL
Rob W. M. Van Soest et al/W: 123TR
Scripps, Lisa Levin/W: 127BR
St Petersburg Paleontological Laboratory: 96
Stef Kriel: 23, 213
Teresa Martin: 23 misc, 119
Terry Keenan/Steve Etches Collection: 91BL
Tim Evanson/W: 126TL
Tim Vickers/W: 36
Tom Kapitany 8/9, 30/31, 32, 33BR, 104, 113, 142BR, 185TR
University of Kansas, NHM/W: 170
US Fish and Wildlife Service/W: 166CL, 185BR
Utagawa Kuniyoshi/W: 56
Visarute Angkatavanich 13, 173BL
Woods Hole Oceanographic Institution, Jack Cook: 197
Zoic srl, Latiaxis srl: 110/111

REFERENCES & FURTHER READING

In tune with today's digital realities we have provided a complete set of references and suggestions for further reading at the book website www.BeyondExtinction.AtOne.org and this will be constantly updated. You can also participate in the discussion on issues raised by the book on our Facebook group Beyond Extinction – The Eternal Ocean.

CAPTIONS FOR BOOK COVER

Front cover images: A moon jellyfish *Aurelia aurita*, bristling with an attendant over-population of brittle stars *Ophiocnemis marmorata*, Mozambique, Bazaruto Archipelago. Note a tiny juvenile brittle star perched atop the jellyfish bell. Photo by Andrea Marshall.

Back cover images: A variety of reef life – all of which have a fossil history of more than 200 million years. Raja Ampat, Papua Province, Indonesia. Photos by Anton Musgrave.

CAPTIONS FOR THE SLIPCASE IMAGES ON THE COLLECTORS' EDITION

Front: Fossil starfish *Pentasteria longispina* plus two brittle stars. Jurassic, 150 million years old. Weissenstein, Switzerland. Photo by Wolfgang Grulke.

Back (version 1): An endemic rock cod *Epinephelus posteli* at rest against an orange sponge at night. It is included in the IUCN Red List of Threatened Species. Sodwana Bay, South Africa. Photo by Wolfgang Grulke.

Back (version 2): A living pleurotomariid gastropod *Perotrochus quoyanus*. Collected at 215 m depth. Roatan, Honduras. This family is the oldest gastropod lineage with a fossil history going back more than 500 million years. They were considered extinct until a live specimen was found in 1879. Photo by Charles Rawlings.

THE CONTINUITY OF ANIMAL LIFE

A Conceptual Phylogeny & Family Tree

Originally published in *Beyond Extinction: The Eternal Ocea. Climate Change and the Continuity of Life* 2019 by Wolfgang Grulke

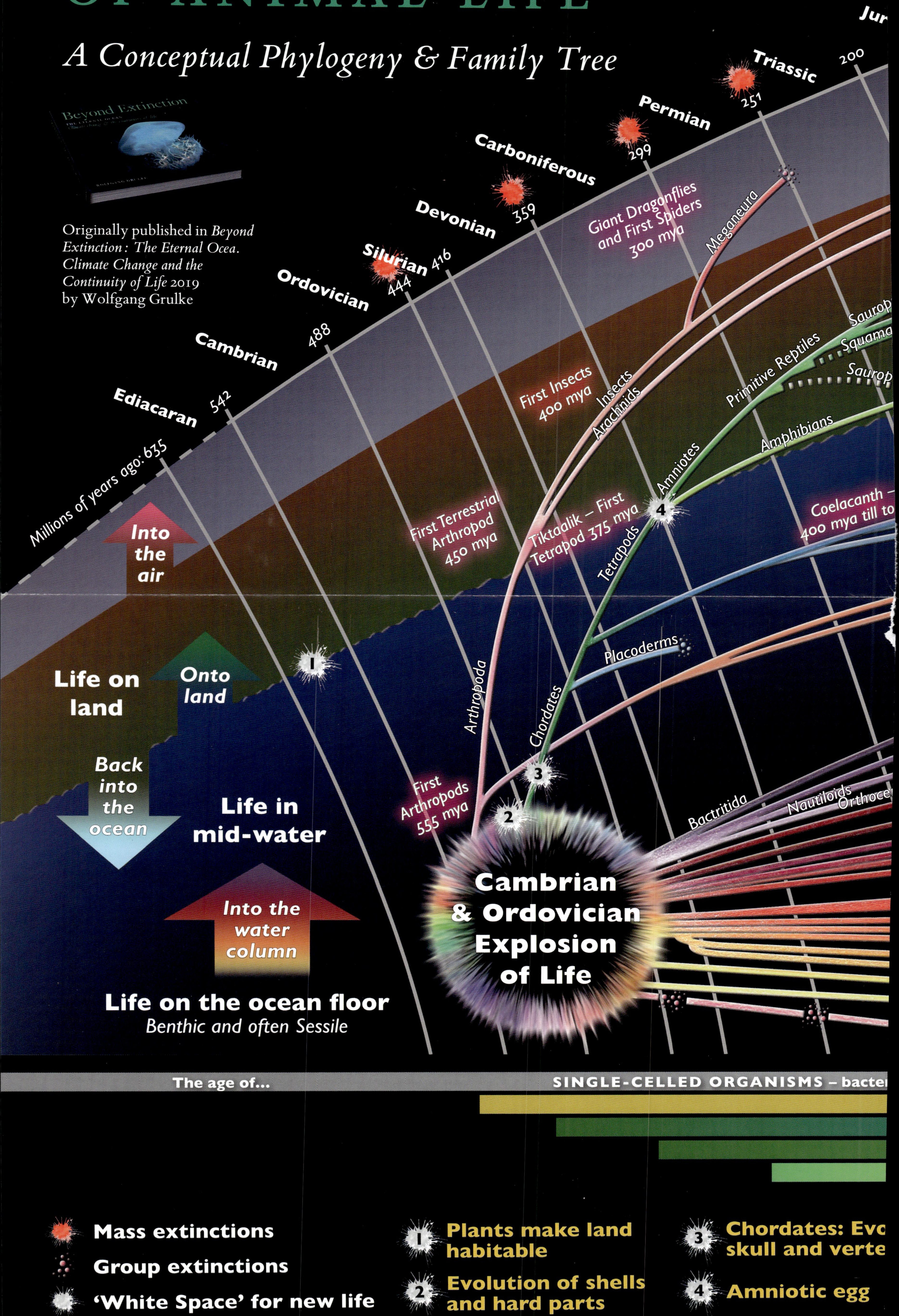

THE CONTINUITY OF ANIMAL LIFE

A Conceptual Phylogeny & Family Tree

Originally published in *Beyond Extinction: The Eternal Ocea. Climate Change and the Continuity of Life* 2019 by Wolfgang Grulke

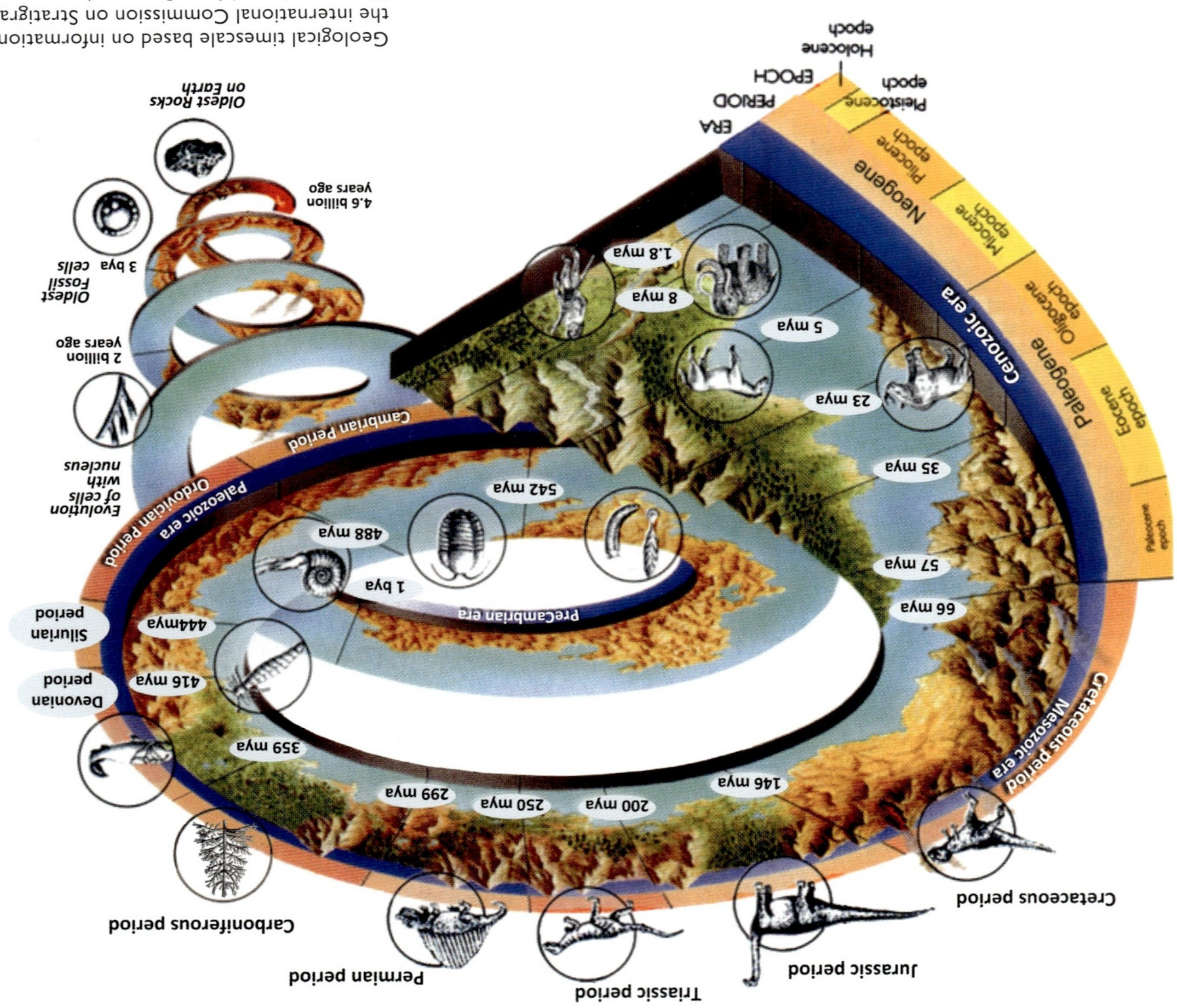

Geological timescale based on information from the international Commission on Stratigraphy Diagram adapted from *Press et al, 2000, Understanding Earth*

These two graphics originally published in *Heteromorph: The Rarest Fossil Ammonites. Nature at its Most Bizarre* 2014 by Wolfgang Grulke

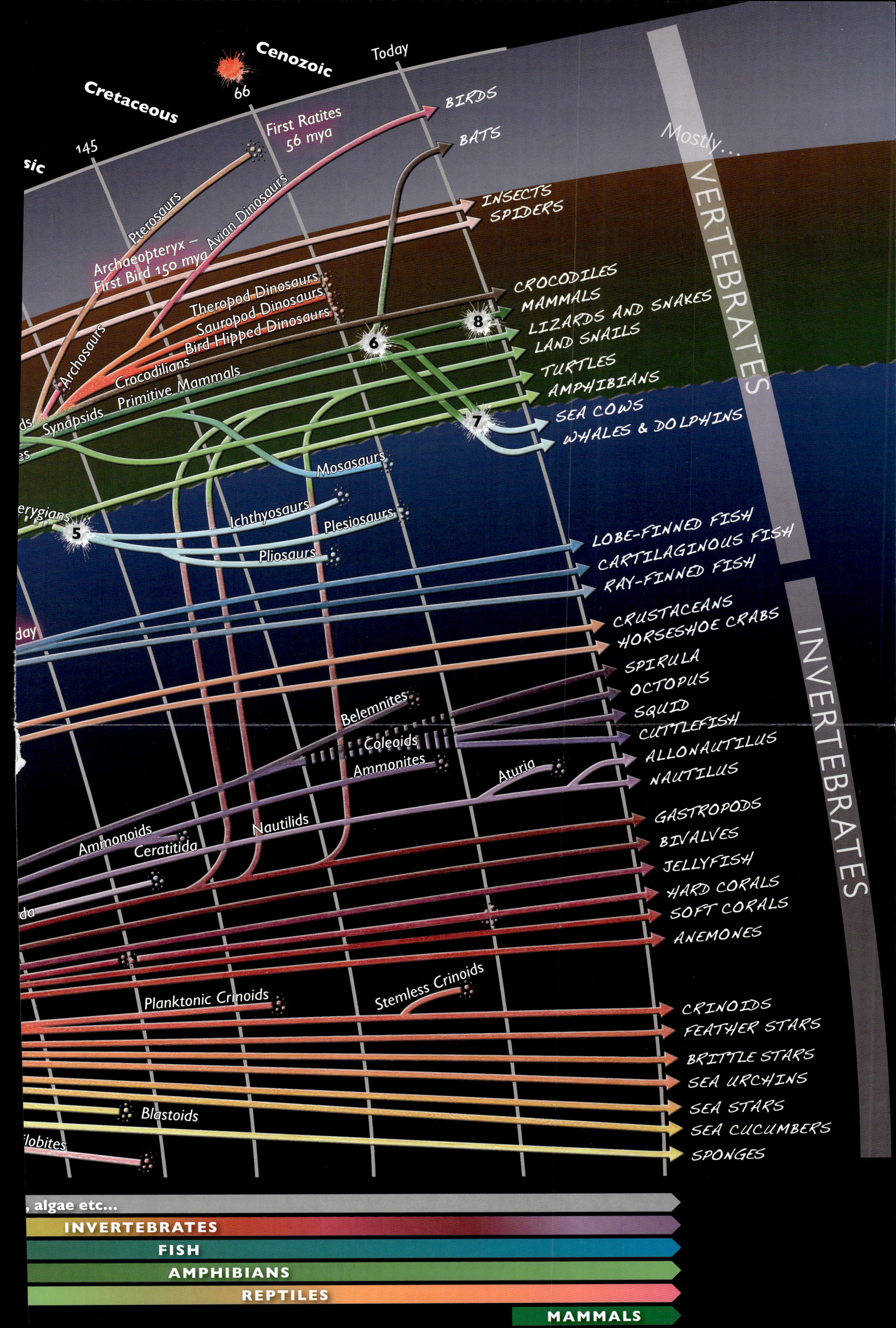

tion of
a

5 Competition on land. Abundance of fish

6 Absence of predators enabled mammals to thrive

7 Abundance of fish encourages mammals to enter the sea

8 *Homo sapiens* emerges

An Epoch-based Geologic Time Scale and Mass Extinctions

Era	Period	Major events
Cenozoic	Quaternary (1.8 mya)	Evolution of modern humans
Cenozoic	Tertiary (66 mya)	Mammals and birds diversify into space left by dinosaurs
Mesozoic	Cretaceous (146 mya)	Extinction of non-avian dinosaurs Extinction of ammonites First primates First flowering plants Heteromorph ammonites diversify
Mesozoic	Jurassic (200 mya)	Dinosaurs diversify First birds evolve from theropod dinosaurs
Mesozoic	Triassic (250 mya)	First mammals First dinosaurs First true ammonites
Paleozoic	Permian (299 mya)	Reptiles diversify First modern trees: conifers, ginkgoes & cycads.
Paleozoic	Carboniferous (359 mya)	First four-legged vertebrates First reptiles Huge forests & swamps create coal deposits
Paleozoic	Devonian (416 mya)	First insects & spiders Fishes diversify: bony fishes & sharks First amphibians First seed-bearing plants
Paleozoic	Silurian (444 mya)	First vascular land plants
Paleozoic	Ordovician (488 mya)	Diversification of trilobites and cephalopods Barren continents
Paleozoic	Cambrian (542 mya)	First trilobites First primitive molluscs & cephalopods First jellyfish, echinoderms & primitive fish
Late Proterozoic	Pre-Cambrian (650 mya)	First soft-bodied metazoans First skeletal elements Stromatolites First traces of animals